Deep Time

Published in 2021 by Welbeck

An imprint of Welbeck Non-Fiction Limited, part of
Welbeck Publishing Group
20 Mortimer Street
London W1T 3JW

A catalogue record for this book is available from the British Library.

ISBN: 978 1 78739 743 9

10 9 8 7 6 5 4 3 2 1

Printed in Dubai

Editorial: Tall Tree Books / Isabel Wilkinson
Design: Russell Knowles / Emma Wicks
Picture Manager: Steve Behan
Production: Marion Storz

Deep Time

A journey through 4.5 billion
years of our planet

Riley Black

WELBECK

Contents

Deep Timeline ... 6

Deep Timeline

13.77 bya
The Big Bang: Time and space begin with a colossal expansion from a point of infinite density, known as a singularity.

13.77 bya + 1 second
Decoupling: Changes in the density of the newborn universe allow neutrinos, almost massless 'ghost' particles, to roam free for the rest of time.

13,770,380,000 years ago
First radiation: The universe becomes transparent to light; the radiation from this moment can still be seen today in the form of the Cosmic Microwave Background.

13.52 bya
Era of star and galaxy formation: The first stars and galaxies begin to form as clouds of gas collapse under the influence of gravity, igniting fusion and blazing into life.

4.56 bya
Formation of the solar system: The centre of a huge disc of gas and dust collapses to form our star, while around it smaller aggregations coalesce.

4.55 bya
Formation of the Earth: A ball of rock and dust aggregates to form the young planet.

4.51 bya
Formation of the Moon: A small planetoid smashes into the baby Earth, disintegrating into a cloud of rock and dust that coalesces to form the Moon.

4.5 bya
Moon lava: Basaltic flows on the Moon create the oldest rocks that can be seen there today.

4.49 bya
Fire from the sky: Meteorites bombard the young Earth, bringing water and potentially leaving traces of rock that is older than the Earth.

4.49 bya
Hell on Earth: The Hadean Eon begins, so named because most of the Earth is a hellish inferno of molten rock for most of this period. Incredibly some fossils suggest that life exists right in the middle of this era, around deep-sea vents.

3.9 bya
Start of the Archaean Eon: The Earth's oldest rocks date to around now, as does the first bacterial life, which left structures known as stromatolites. The atmosphere at this time is a toxic mix of methane and ammonia.

2.5 bya
Start of the Proterozoic Eon: Many of the greatest events in Earth's biological history take place in this eon, which lasts until 542 mya.

2.5 bya
Start of the Great Oxygenation: Photosynthetic bacteria start to produce oxygen, beginning a process of atmospheric transformation known as the Great Oxygenation Event.

2.45 bya
Snowball Earth: Oxygen reacts with methane in the atmosphere to produce carbon dioxide, which has a much weaker greenhouse effect. The Earth is plunged into a global glaciation for hundreds of millions of years.

1.5 bya
Complex cells: First eukaryotic cells are created when endosymbiosis leads to origin of mitochondria.

730 mya
Avalonia: Formation of the first rocks of what will become Britain, with the creation of an ancient geological formation known as the Avalonia terrane.

575 mya
Strange creatures: A sudden flowering of evolution leaves a trail of weird fossils in the Ediacara Hills of Australia and elsewhere.

541 mya
Cambrian Explosion: All of the branches of modern multicellular life appear overnight, in geological terms, near the start of the Cambrian Period.

488 mya
Ordovician Period: In this period most of the land mass is gathered together into the Gondwana supercontinent, where the ecology is dominated by the marine ecosystem of shallow seas. The first bony fish appear and the first organisms begin to colonize the land.

c.400 mya
Creepy-crawlies: During the Devonian Period, the first insects evolved.

c.330 mya
Black stuff: The waterlogged forests and swamps of the Carboniferous Period lay down vast deposits of peat, which will eventually form into coal.

c.275 mya
Permian Period: The supercontinent Pangaea contains most of the landmass, and mammal-reptile hybrids known as synapsids dominate terrestrial ecosystems.

251 mya

Extinction: The greatest mass extinction of all time wipes out 90 per cent of species and clears the way for new life forms to dominate on land and in the sea. The cause may have been the colossal volcanic eruptions that formed the Siberian Traps.

c.200 mya

Age of the dinosaurs: From the Triassic Period onwards, dinosaurs dominate the land while ammonites thrive in the seas.

c.180 mya

The break-up: The Americas split from Africa and begin their long journey west.

c.175 mya

Jurassic park: The Jurassic Period sees the evolution of massive dinosaurs, the first birds and the formation of North Sea oil.

c.125 mya

Life in bloom: The first flowering plants appear around this time.

66 mya

Impact: A huge asteroid slams into the coast of Mexico, wiping out the dinosaurs and many other life forms and marking the transition between the Cretaceous and Tertiary Periods.

c.50 mya

Pile-up!: The Indian landmass collides with the Asian landmass, pushing up a huge mountain range – the Himalayas.

c.6 mya

Forebears: Hominids evolve in Africa, with early species such as *Ardipithecus*.

c.300,000 years ago

Walking tall: Anatomically modern humans evolve in the East African Rift Valley. They have a sophisticated tool set and can alter local ecology through burning.

c.100,000 years ago

Out of Africa: Successive waves of migration of anatomically modern humans leave Africa, initially spreading through the Near and Middle East and into Asia and southern Europe.

c.60,000 years ago

Land down under: Humans reach Australia, although the exact date of arrival is contested.

c.13,000 years ago

Coming to America: The first humans reach the Americas from Siberia, probably following the coast to skirt the massive ice sheets that still blanket the north.

c.10,500 years ago

Neolithic Revolution: Humans start to domesticate plants and animals, marking the start of farming.

c.10,000 years ago

Defrost: The end of the last Ice Age sees the retreat of the ice caps to their last redoubts in Antarctica and Greenland.

8,000 years ago

Under the wave: The area between ancient Europe and England called Doggerland is flooded by a tsunami.

c.5,000 years ago

Tablet time: The first written records mark the end of prehistory and the beginning of recorded history, as the first civilisations arise in Mesopotamia.

c.4,789 years ago

Methuselah germinates: The oldest tree still alive today germinates in the White Mountains, California.

c.4,500 years ago

The Great Pyramid: Ancient Egyptian Pharaohs rule over the whole Nile Valley and construct great monuments. Meanwhile in Britain, neolithic peoples are constructing Stonehenge.

c.4,000 years ago

The last mammoth: Woolly mammoths may have survived until now on Wrangel Island in Siberia.

c.3600 years ago

Cataclysm: The eruption of Thera in the Mediterranean devastates the region and helps to end the Minoan civilization.

c.3000 years ago

Iron Age: Ancient peoples master metal working and start to make tools, armour and weapons of iron.

1,893 years ago

Another brick in the wall: Romans in England complete Hadrian's Wall.

What is time?

Our lives are ruled by time. There are time spans for our lives, our expectations and what we wish for the future, all fractions of a history that stretches billions upon billions of years into the past. In fact, time seems so ever-present that it's easy to take it for granted. But time as we know it did not have to exist as it does for us. At the beginning of our journey into Deep Time, it's first worth asking what time *is*.

The glib answer to what time is might come from your watch, computer or phone. But that's simply a way we count time rather than time itself. Looked at one way, time is a series of events stretching from the past through the present and into the future. Change and difference are inherent to time itself. If nothing ever changed, the universe would truly be timeless.

Then again, we might take a more technical stab at the question. Time is actually a dimension – a part of what makes up our universe – that allows an object in space to be in more than one position in that space. Think of the last time you walked from one end of a street to the other as you were on your way somewhere. The fact that you were able to exist at the start of the street and move along the pavement to a totally different position in space, on a rotating planet, moving through its orbit in a solar system, spiralling in a galaxy, is a sign that we live in a universe with time.

The arrow of time

That's not to say that this is all there is to time. It's just a starting point, the barest minimum that then opens up to other possibilities. One of the fundamental aspects of time is that it is directional. Time is not a cycle where the past becomes the present and then circles back around. Our universe is shaped by time's arrow, the movement from the *was* to the *is* to the many variations of the unknowable

Above Due to the arrow of time, we often see cups shatter, but we never see shattered pieces spontaneously assemble themselves into a cup.

Opposite GPS satellites need to correct their clocks to allow for the warping of space-time caused by the Earth's gravitational pull.

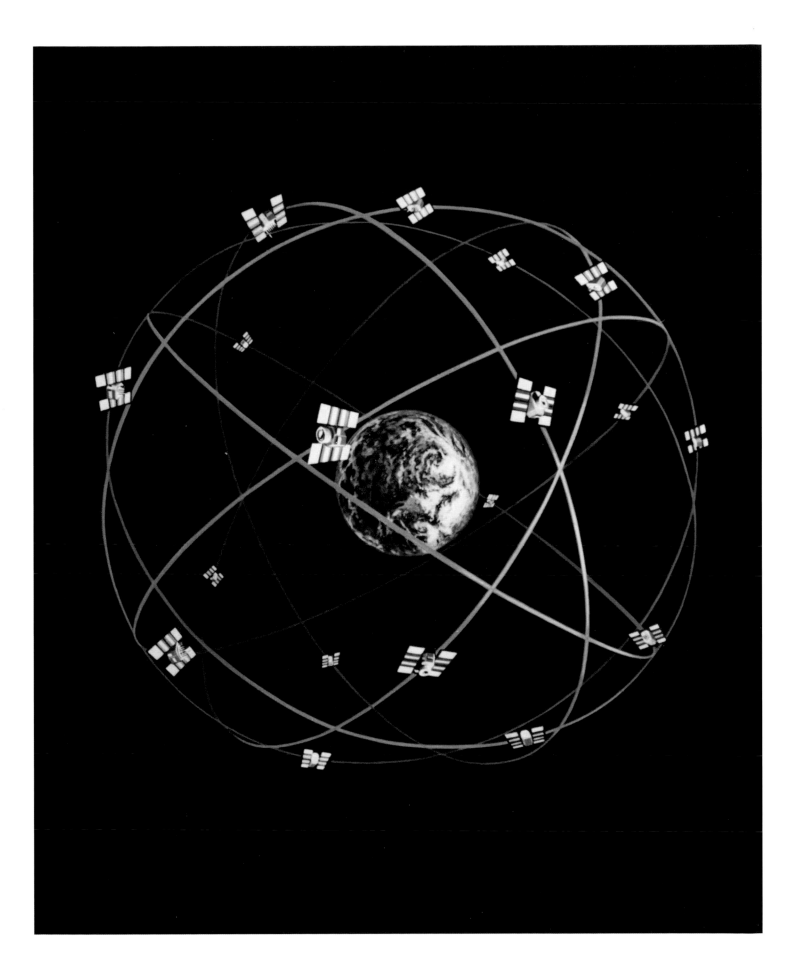

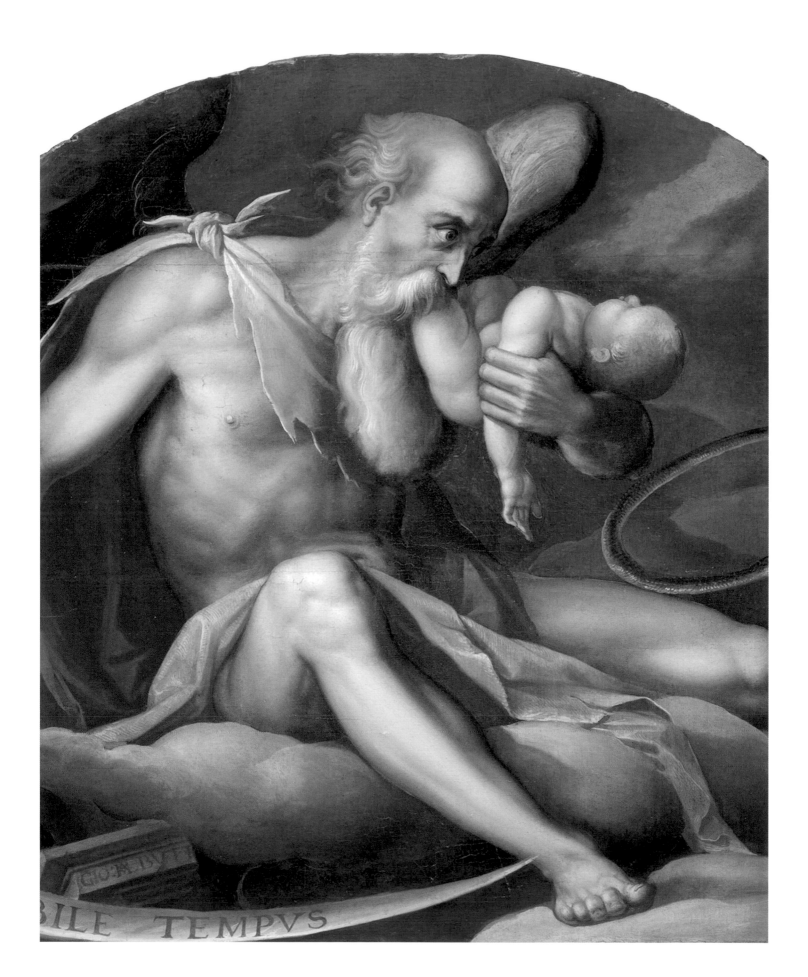

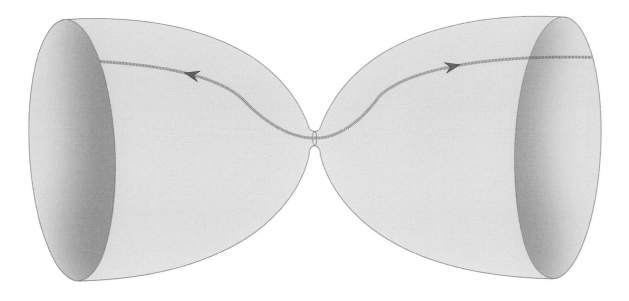

future, always in a single direction. We are at the tip of that arrow, with moments of the past as inaccessible as the future.

Another way to think about time is through the phenomenon of entropy, or a measure of what might be called disorder. If you make a sandwich and leave that sandwich on the counter, even in a room with no hungry pets to eat it, microorganisms and other processes will eventually break down your lunch and turn it into something very different. The universe, in a sense, is much the same. Entropy has been increasing since the first moment of time, and that is an inescapable fact of life.

Figuring out why our universe started with relatively low entropy but experiences increasing entropy over time is a task that continues to confound physicists and cosmologists, but it might be a clue to something else. We often speak of the Big Bang as if it came from absolutely nowhere and was the beginning of everything. But that's an assumption. If the Big Bang emerged from something earlier, then those earlier conditions might have influenced what our early universe was like and how time would play out as we experience it.

Other universes?

Much remains hypothetical, but it's possible that our universe began as part of another that did not have directional time as we do. Without the arrow of time, this predecessor would have had time that was impossible to perceive because there would be no before or after. But if the Big Bang started by splitting off from such an existing universe, that might explain why our universe went through an earlier period of low entropy and increases in entropy over time. The beginning might not be the ultimate beginning but a starting state that carried certain conditions with it.

Even as our understanding of what time is has changed and brought up new questions, though, we know time exists because we perceive it. We feel the passing of time in our own bodies and see it with our own eyes. That's part of why we wish to keep track of it all. Long before our present era, people kept track of time by digging pits to track lunar phases, created sundials, invented calendars, and even named deities – like the ancient Greek god Chronos – to try to understand this fact of nature. Where time came from is still difficult to grasp, held in a past that's difficult to see, but we can at least get back nearly to the moment when our universe's clock started ticking. Even if there is an older beginning, there's hardly a better place to start than with the Big Bang.

Above British physicist Julian Barbour has suggested that the Big Bang represents a 'Janus Point' from which time heads off in two directions, one of which we experience as our universe.

Opposite The Greek god Chronos is often depicted with a scythe, reflecting the destructive nature of time.

Neutrinos

Just how deep is time? We now have increasingly precise answers to this question. According to the latest measurements, our universe started 13.77 billion years ago, give or take about 40 million years, with the Big Bang. This was the moment at which both space and time came into existence. The moment itself remains shrouded in mystery, but we are surrounded by relics from the second after the Big Bang. These are neutrinos.

The moment of the Big Bang is theorized to have occurred as a singularity – a dimensionless point containing all the energy needed to make a universe. If you find the concept of a singularity difficult to grasp, don't worry. You're in good company. A gravitational singularity (or space-time singularity) is a location where the quantities that are used to measure the gravitational field become infinite. Currently, physics as we know it only starts to make sense a tiny fraction of a second after the singularity, following a period of rapid expansion known as cosmic inflation. Inflation is thought to have occurred between 10^{-36} seconds and 10^{-32} seconds after the Big Bang. In this time, the universe doubled in size at least 85 times, reaching a volume somewhere between that of a grain of sand and a football. A moment later, the first neutrinos were created.

Neutrinos are very tiny subatomic particles with nearly zero mass. Physicists calculate that neutrinos have maybe a millionth as much mass as an electron. And they're everywhere. The universe is full of them, and yet, because of their almost immeasurably low energy, neutrinos hardly interact with the matter of the universe at all. This makes them difficult to detect, even as neutrinos are constantly passing through our bodies every second of every day. Their existence was confirmed in 1956 in an experiment dubbed 'Project Poltergeist'. After five years of trying, American physicists Frederick Reines and Clyde Cowan Jnr detected neutrinos by observing their

so-called inverse beta-decay interaction with matter. This was Big Physics, using a nuclear reactor to create streams of neutrinos and detecting them with huge tanks filled with a water and cadmium chloride solution. The tanks had to be buried underground to shield them from cosmic rays that would have messed up the results.

Below Clyde Cowan Jnr (front left) and Frederick Reines (front right) with the Project Poltergeist team in 1955.

Opposite Scientists work on the Baikal Deep Underwater Neutrino Telescope, which was placed at the bottom of Lake Baikal, Russia, in 1990.

Above The IceCube Neutrino Observatory at the South Pole.

Cosmic Neutrino Background

Now that we have found them, neutrinos offer researchers an opportunity to delve into Deep Time. Many of the neutrinos around today are left over from the earliest moments of the universe, almost-indiscernible time capsules that record the conditions in the first second after the Big Bang. The key piece of evidence comes from what researchers call the Cosmic Neutrino Background.

In the moment of the Big Bang, the heat of the early universe created a huge amount of electromagnetic radiation as well as a massive quantity of neutrinos and their antimatter counterparts called antineutrinos. The number truly is astronomical. Pick any cubic metre of space in the universe and within it there will be about 100 million neutrinos. Some of these were more recently created in high-energy events like supernovae, but many are nearly as old as the universe itself. These are known as relic neutrinos.

Given their low energy and minuscule mass, neutrinos mostly keep to themselves. They don't have a charge, they don't emit light, and, as their name suggests, they often seem rather neutral. Physicists estimate that the Cosmic Neutrino Background has a temperature of about minus 271 degrees Celsius (2 Kelvin – two degrees above absolute zero), and a huge input of energy is needed to warm them up to a state where they begin to interact with matter. Sometimes neutrinos can pick up energy from radiation inside the sun or in supernovae, but for the most part, they are just part of the fabric of the universe. And this is precisely what has inadvertently preserved neutrinos since the moment after the Big Bang.

Ending interactions

While the Big Bang released an incredible amount of heat and energy, the rapid expansion of our universe caused temperatures to plummet very quickly. Almost as soon as they were created, neutrinos were deprived of the massive energy input required to allow interactions with matter. In a snap of the fingers after the Big Bang, in the very first second of the universe, there was no longer enough heat and energy for the just-created neutrinos to do much of anything. Since neutrinos don't decay or transform into other particles, they've persisted all this time. There may not be anything else in the universe that is as long-lived.

Detecting neutrinos in their typical, low-energy state has proved nearly impossible, but scientists have been able to identify their presence by looking at how they have affected other parts of the universe. The cool temperature of the universe's neutrinos affects the Cosmic Microwave Background, the faint background radiation that fills our universe. By looking at these clues, the dramatic changes that happened in the first second of the universe's existence begin to come into focus.

Baryon acoustic oscillations

While it may be difficult to come up with the standard measure of all things,
the universe itself has something that scientists call a standard ruler. This is
a way to understand and compare distance in the universe.

The standard ruler has a length of 490 million light years. While that might seem like an awfully long ruler, consider how big the universe is. Physicists estimate that the observable universe is about 93 billion light years in diameter, which is over 189 standard rulers wide.

Scientists have settled upon a length for the standard ruler through studying a phenomenon called baryon acoustic oscillations. A baryon is a composite subatomic particle made up of at least three quarks. The most familiar baryons to us are protons and neutrons, each made of three quarks, which form most of the mass of visible matter.

Conservation of particles

Owing to the way that matter and energy are conserved, the number of particles in the universe has been the same ever since the Big Bang. However, the behaviour of these particles was changed by the expanding universe, which created a greater volume and therefore caused matter to be less dense within it. The way this expansion altered the behaviour of particles left literal ripples in space.

Think back to the time soon after the Big Bang began our universe's expansion. Following the Big Bang and the rapid cooldown that caused neutrinos to become so neutral in the first second of existence, the universe was made up of an extraordinarily hot and dense plasma that was brimming with subatomic particles such as electrons, protons and neutrons.

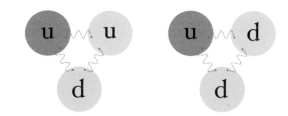

Charged plasma

Plasma is a tricky substance to categorize. It is a state of matter similar to a gas, but in plasma, some or even all of the electrons have been winnowed away such that positively charged ions can move freely. Effectively, plasma is a charged gas and, despite its rarity on Earth, it may be the most common state of matter in the entire universe. This was true during the earliest phases of our universe, too, when primordial plasma created a vast sea of charged ions in space. Rather than being even, however, some spots in the primordial plasma were much denser than others, and this was the key to the production of what were essentially shock waves in space.

Above Protons (left) and neutrons (right) are baryons made of three quarks.

Opposite Top The Cosmic Microwave Background fills space. It was emitted at the time the baryon acoustic oscillations were produced.

Opposite Bottom Pairs of galaxies are commonly separated by one length of the standard ruler.

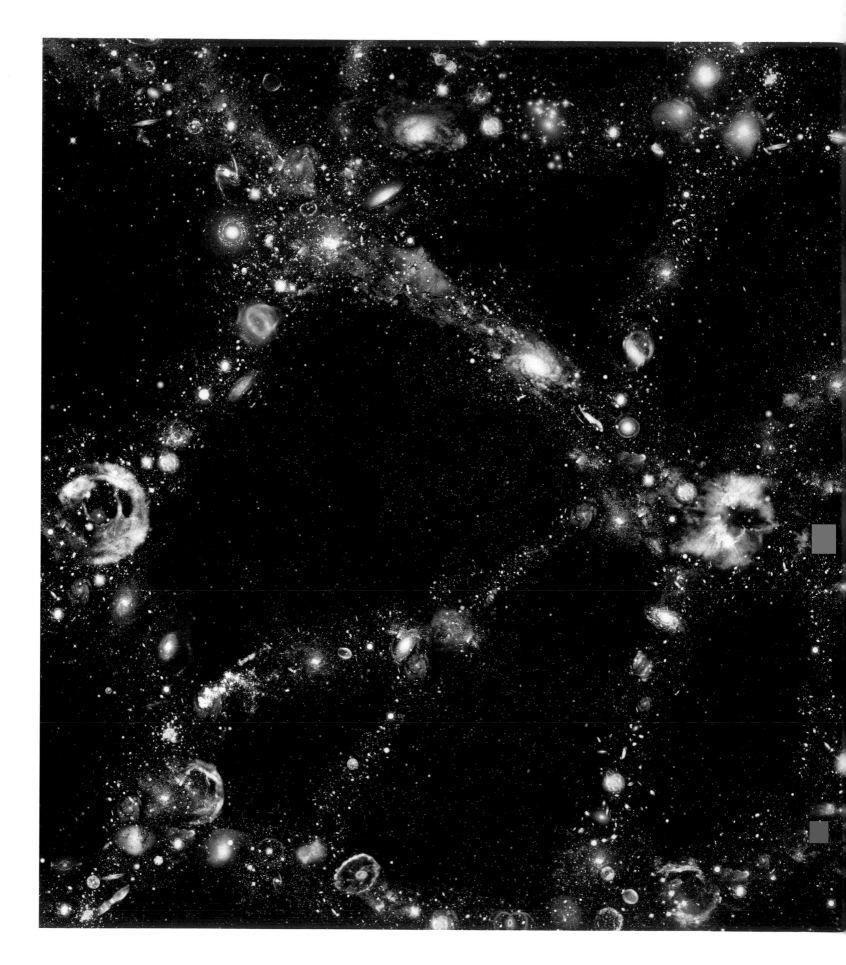

Agglomerating pockets of matter in space gain gravity, pulling more matter towards them. This is how many of the planets and other celestial bodies eventually formed, with clumps of matter attracting more matter to themselves through time. During the early days of the universe, however, this form of interaction was more complicated. As gravity pulled on accumulating, dense clumps of matter to bring more particles together, it also generated heat. The heat, in the form of photons, created a pressure that pushed matter apart. When that matter cooled, gravity once again took hold and started to pull more matter together. This cycle kept repeating itself as the universe continued the early phases of its expansion.

Acoustic waves

This constant push and pull, which occurred about 370,000 years after the Big Bang, created waves through the universe. These are the phenomena that scientists have called baryon acoustic oscillations, which eventually set the length of the standard ruler. The waves in the primordial plasma were similar in structure to sound waves, which are longitudinal waves in which particles oscillate along a direction parallel to the direction of propagation. Imagine protogalaxies scattered along the peaks and troughs of a wave. Eventually, gravity started to win out and the universe's first electrically neutral atoms began to form from the plasma. Matter could now cool more quickly, which meant that gravity suddenly had an edge.

While they happened well over 13.7 billion years ago, the pattern of the baryon acoustic oscillations can still be seen today. The waves were like a set of cosmic wrinkles that determined where galaxies would form in the universe. In fact, part of the evidence for these oscillations comes from the way that galaxies seem to cluster along lengths predicted by such interactions. The standard ruler is defined as the maximum distance the acoustic waves could travel before atoms could form, and pairs of galaxies seem to be separated by this distance more than by other lengths. Their distribution isn't random, in other words, but is informed by the early perturbations in the primordial plasma. Armed with this knowledge, researchers have a means to calculate the rate of retreat of pairs of galaxies, whose separation appears to get smaller as they move further away. These measurements are used to calculate how fast the universe is expanding, shedding light on the nature of dark energy, the mysterious force that drives cosmic expansion.

Left The distribution of galaxies across the universe is not random. They form clusters whose positions are predicted by the theory that they were determined by the baryon acoustic oscillations.

The Hubble Deep Field

The night sky is a strange place to contemplate, Deep Time-wise. We stare out from our earthly vantage point, feet planted in the here and now, but what we're looking at is a stream of time. When researchers use high-powered telescopes such as the Hubble Space Telescope to look even further beyond, they can see even further back in time.

We don't see all those little lights – the distant stars, planets and galaxies – as they are now, but rather as they were at some point in the past. We see light that may have been travelling for immense amounts of time.

A view from space

The Hubble Space Telescope looks at such vistas with far more clarity than we can. Launched in 1990, the Hubble was the first major optical telescope to be launched into space. The ability to look through lenses and take pictures from space has been a huge boon. Unlike observatories here on Earth, the Hubble doesn't have to deal with light pollution from cities or the way the atmosphere can distort images. This prime vantage point has resulted in breathtaking views of bodies in space that might not look like any more than a pinprick to the naked eye, if they'd be visible at all. Among the most spectacular images captured by the Hubble is the HDF – the Hubble Deep Field.

Between 18 December and 28 December 1995, the Hubble Space Telescope focused in on a small part of the constellation Ursa Major. Technicians took 342 photographs of the area, then stitched them together to create an image dotted with stars and galaxies. This composite image represents about one 24-millionth of the entire night sky. Measured in telescopic terms, it is 2.6 arcminutes long on each of its four sides. Altogether, there are 3,000 visible objects within the Hubble Deep Field, including some of the most distant galaxies in the universe.

The galaxies visible in the image came in different shapes. Some were known as irregular galaxies, or galaxies that don't have any regular shape to them. Others were spiral galaxies, taking the shape of a flat, rotating disc around a central point. But that wasn't all. About 50 of the galaxies in the photo seemed to have little blue objects near them. These blue dots are places where new stars are forming, old stars that have run out of fuel to become white dwarfs and quasars.

Measuring redshift

As scientists researched the contents of the Hubble Deep Field, they found that many of the visible galaxies had high redshift values. The redshift of a celestial object like a galaxy is a lengthening in the wavelength of its radiation that happens as it travels through space. In our expanding universe, more distant objects have higher redshift values. It is called redshift as red is the colour in the visible spectrum with the longest wavelength, but in reality, the radiation may be well outside the visible spectrum.

Opposite The Hubble Space Telescope has made more than 1.4 million observations since its launch in 1990. It takes images that range from ultraviolet through the visible spectrum into the near-infrared.

Above The eXtreme Deep Field, taken between 2002 and 2012.

Opposite The Hubble Deep Field, taken in 1995.

The opposite effect – a shortening of wavelength – is called blueshift. The fact that the expansion of the universe causes more distant objects to move away from our observation points on Earth more quickly than nearer objects compounds the effect. Using these principles, researchers calculated that the most distant visible galaxies were about 12 billion light years away.

Looking deeper

The Hubble Deep Field was the first clear look at how galaxies form, not only seeing far but also providing more information about how our universe came to be. It was so successful, in fact, that researchers launched two additional projects to compare the original Hubble Deep Field with other vantage points. Between 24 September 2003 and 16 January 2004, the Hubble focused on a different area of space to create an image called the Hubble Ultra-Deep Field, which contains images of about 10,000 galaxies. Despite the greater number of visible galaxies, though, the region looked very similar to the original HDF image. Upon analysis, researchers took this as evidence that the universe is largely uniform over vast distances and that the Earth exists in a typical part of the universe – an idea called the cosmological principle, which predicts that the distribution of matter in the universe will appear essentially even when looked at from a broad enough scale.

Researchers kept snapping new images through the years to make even more comparisons. In 2012, experts released the Hubble eXtreme Deep Field, a zoomed-in part of a previous Hubble Deep Field image. The image was extremely detailed, containing about 23 days' worth of exposures taken over 10 years and covering one 32-millionth of the sky. This new image contained images of about 5,500 galaxies, including some that were about 13.2 billion years old – distant galaxies that formed about half a billion years after the Big Bang, among the oldest known. All those colourful dots against the black are a time capsule of the ancient universe, visible through the researchers' new eye in the sky.

Supernova 1997ff

In the whole of space, there is no bigger explosion than a supernova. It is an event that is millions of years in the making, although its most violent phase may last a matter of seconds. Evidence from an extremely ancient supernova has shed light on the nature of the early universe.

Supernovae can be triggered in two ways. Sometimes two stars are relatively close together, both orbiting the same point in space. In these situations, one of the stars may expend its fuel a little faster than the other. That star becomes a white dwarf – essentially a dead star, which starts to attract matter from its stellar companion. But more matter means more gravity, which only increases the accumulation, until the excess makes the white dwarf explode into a supernova.

The other way to create a supernova involves a similar process but within a single star. When a star is out of fuel and begins to die, its core gains mass. Among more massive stars, the centre of the star becomes denser and heavier to the point where the core collapses and triggers a supernova.

The senescence and death of stars can take millions of years. Then, in less than a second, the core collapses to begin a shock wave that can take hours to reach the surface. This destruction creates a glow that ramps up over the course of months and may last as long as a few years. From our vantage point on Earth, though, we're often lucky if we can spot a supernova for even a brief moment.

A new star?

The most recent supernova to appear in our night sky, visible to the naked eye, was Kepler's Supernova. During 8 and 9 October 1604, people across the Northern Hemisphere noticed a very bright star in the part of the sky that includes the constellation Ophiuchus, 'the snake-handler'. At the time it was spotted, the supernova was brighter than any other star.

The event certainly made an impression on the German astronomer Johannes Kepler, who described the supernova in his book *De Stella Nova*, and after whom it was named. So far as researchers have been able to tell, Kepler's Supernova was no further away than 20,000 light years from Earth and was the most recent supernova to be observed within our galaxy, shining in the sky for only two days more than 400 years ago.

Kepler thought he was looking at a new star, hence the name of his book describing it. But this was a moment of destruction in our galaxy, the effects of which are still visible today. In 2020, NASA astronomers were able to spot debris from the supernova that is still moving away from the site of the explosion at speeds of more than 30 million kilometres per hour. That figure alone speaks to how impressive these explosions are.

Opposite A composite false-colour image of the remnant of Kepler's Supernova, made by combining images from the Hubble Space Telescope, the Spitzer Space Telescope and the Chandra X-ray Observatory.

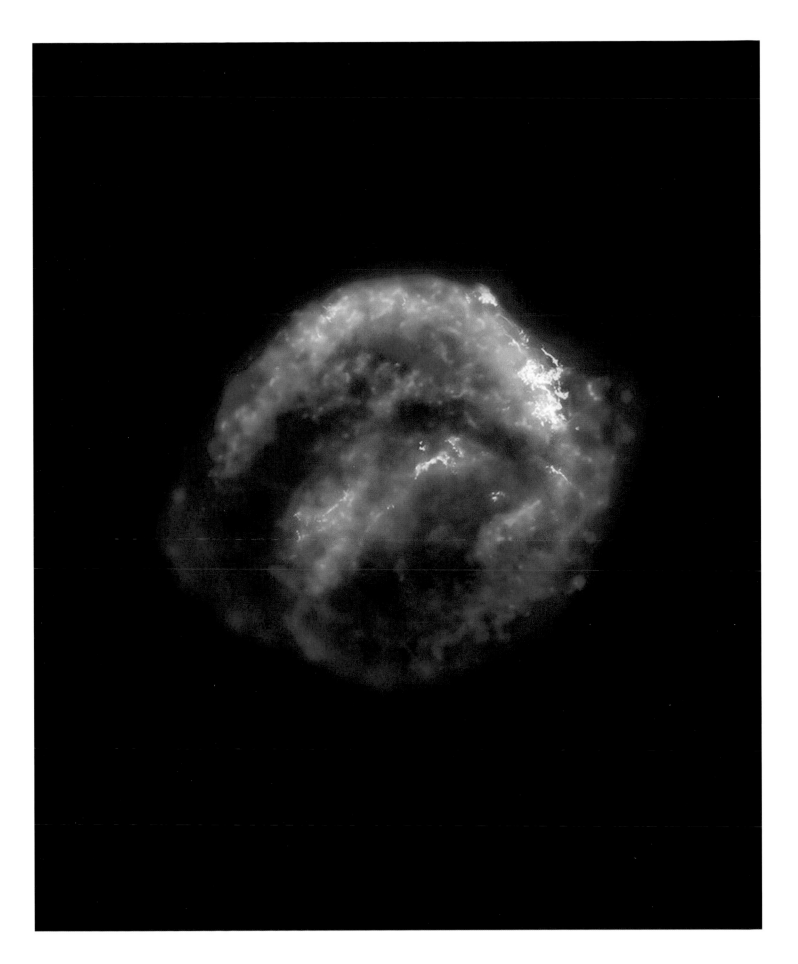

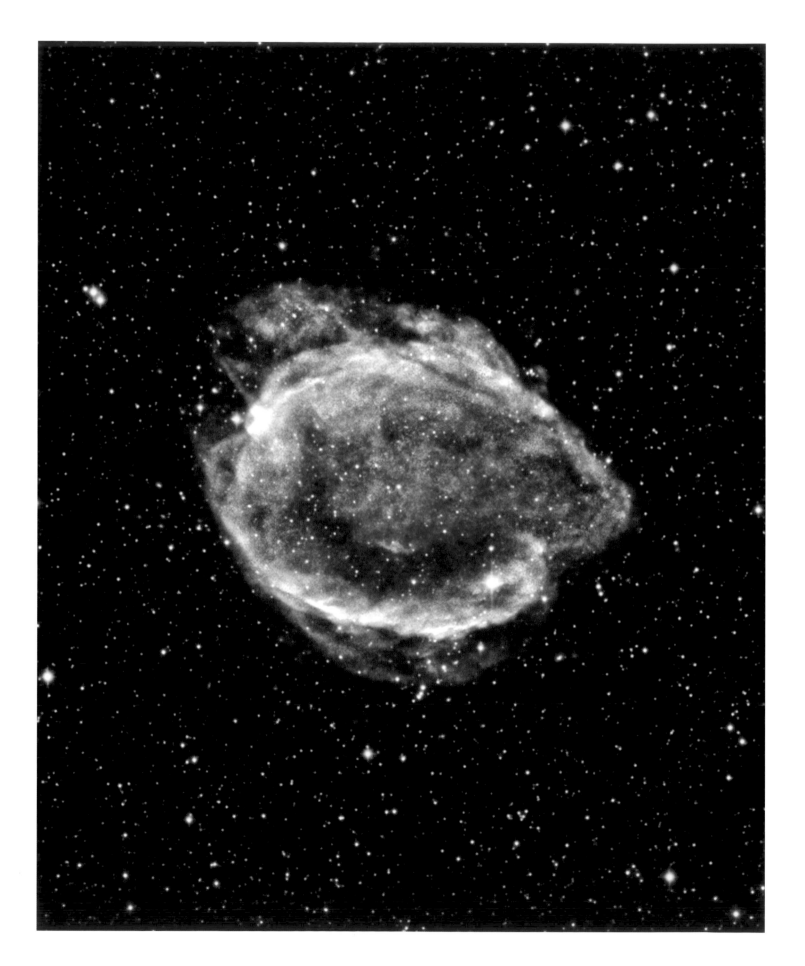

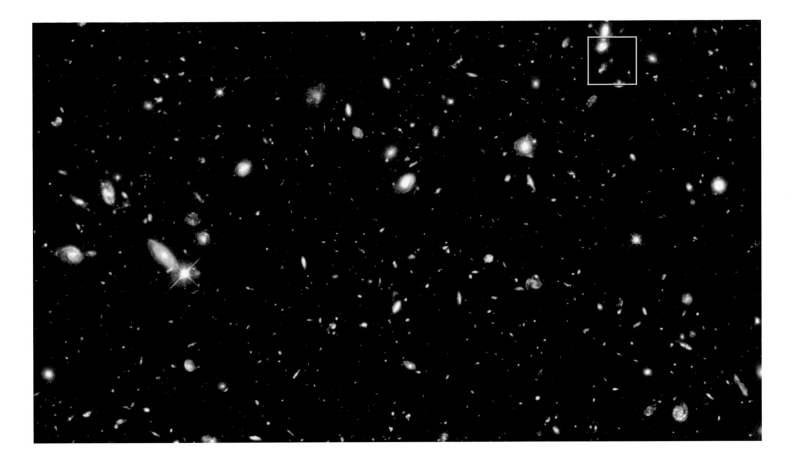

Ancient explosion

For astronomers, supernovae are far more than just beautiful interstellar light shows. These immense explosions can also provide insights into what happened in the ancient universe. One such burst is Supernova 1997ff.

This particular explosion was found by comparing the original Hubble Deep Field images with a follow-up from 1997. Among the various dots in the image, astronomers detected a supernova in a very distant and faint galaxy. Look carefully and you might be able to pick it out as a tiny red smudge towards the top right of the Hubble Deep Field. It was so far away, in fact, that experts had to correct for the time dilation that space can cause at such distances. Two different research teams viewed the same supernova 25 days apart, but for the supernova only nine days passed between observations.

Working from what's known about the expansion of the universe and the distance of the objects, the researchers were able to determine that the supernova popped about 11 billion years ago, making it one of the oldest supernovae ever seen.

Changing rate of expansion

The distance and nature of Supernova 1997ff confirmed that more distant objects in the universe move away from the Earth more quickly, underscoring the idea that we occupy a universe whose expansion is accelerating. The supernova is something of a time capsule. When the star exploded more than 11 billion years ago, the universe was smaller than it is today. That means it was also denser, meaning that the influence of gravity in the early universe was a little bit different, too. By comparing the details of Supernova 1997ff to others that have occurred more recently, astronomers have posited that the great acceleration in the expansion of the universe didn't kick in until it was about 7 billion years old.

What changed? The answer may be related to dark energy. The dense nature of the early universe allowed gravity to keep stars and galaxies closer to each other. But dark energy can overcome gravity, and this force is what drives the sped-up rate of expansion in the latter half of the universe's history. Supernova 1997ff is not just from a very ancient time, but from a distant era when the universe behaved in a very different way. That little red dot is practically a fossil star, a remnant of a time quite unlike our own.

Above Supernova 1997ff is the red smudge in the middle of the highlighted square in the Hubble Deep Field.

Opposite NASA's Chandra X-ray Observatory has taken images of many supernova remnants, including this one of G299, a remnant located within the Milky Way.

The Wold Cottage meteorite

Much of what we know about the formation of the universe, galaxies, and even our solar system comes from our observations of distant space. But sometimes space comes to visit us instead.

On 13 December 1795, a chunk of rock plummeted through Earth's atmosphere to bury itself in a field near the village of Wold Newton in Yorkshire, England. It must have come as quite a shock to farmer John Shipley, who was just a few metres away from this piece of rock that came out of nowhere. It buried itself more than 50 centimetres into the soil – an impact forceful enough that Shipley was reportedly pelted with soil and rocks from the crash. The scene was like something out of a science fiction novel, and the stone was still warm in its miniature smoking crater as people gathered to see what all the hubbub was about.

Witness accounts

Writer Edward Topham lived nearby and took statements from at least three witnesses to the event. Each described a dark object flying through the air before hitting the ground. This had to be the stone that landed by Shipley, and even though thunder was heard on that particular day, those normal weather conditions couldn't explain such a big chunk of stone thudding into the ground. Topham was so impressed by this singular event that he had a monument commissioned for the site where the stone hit the earth.

At the time, no one had a firm idea of where meteorites came from or what they represented. In fact, some naturalists of the time didn't think any matter in space could fall to Earth. One idea was that such rocks were blasted out of volcanoes. This didn't seem reasonable for the stone that fell near Wold Cottage, though, as there were no active volcanoes anywhere nearby. Other experts asserted that these smoking stones were made by lightning, with the intense flash of heat liquefying and kneading together stone into new lumps of transformed rock. But there was no evidence that lightning did anything like this.

Scientific investigation

Wherever the stone that struck the farmland had come from, it was enough of an oddity to become a minor sensation. Topham took initial ownership of the meteorite, displaying it at Piccadilly in London. But he didn't keep it for very long. In 1804, Topham sold the meteorite to English mineralogist James Sowerby for a little more than ten pounds. Sowerby was more interested in the science of the rock than the sensationalism, though. He described the stone in his book *British Mineralogy*. "To introduce

Above Edward Topham (1751–1820).

Opposite A detailed drawing of the Wold Cottage meteorite with its dimensions and weight, made by Topham's daughter, Harriet.

Harriet H. Topham

Hight   30 Inches

Breadth   28 Inches

Weight   56 Pounds

Above The Wold Cottage meteorite is now kept at the Natural History Museum in London.

Opposite An obelisk was erected at the landing site of the meteorite
near the village of Wold Newton in Yorkshire.

a substance, however curious, as having fallen like a meteor from the skies, or as Phaeton from the heavens, might seem absurd," Sowerby wrote. He noted that the rarity of such an object made it important, especially because, as the geologist stated, "we ought, in charity, to wish such may still continue to be rare, as otherwise the consequences might be dreadful."

Sowerby's explanation as to where the rock came from wasn't much more grounded in evidence than those of many others. He proposed that such rocks may be found somewhere in the atmosphere and, in contacting an 'electric fluid', might be shot down to the ground. The idea that this stone had come from space was too much for him. But that is the conclusion that geologists and astronomers would eventually accept. The notion that meteorites fell from space had first been proposed by the German physicist Ernst Chladni in 1794. This idea had been widely dismissed at first, and only started to gain acceptance ten years later after French astronomer and physicist Jean-Baptiste Biot produced an analysis of 3,000 meteorite fragments that fell on the town of L'Aigle in Normandy on 26 April 1803.

Ancient remnant

The Wold Cottage meteorite is the largest meteorite known to have fallen in England. More recent research has indicated that the meteorite is a remnant of our own solar system's formation that's been kicking around space for about 4.6 billion years.

The grey, lumpy stone weighs about 25 kilograms. Experts have categorized it as a chondritic meteorite, meaning that it is made of stone that has not been melted or otherwise altered. In fact, it's a kind of space debris – a fragment of rock that split off and hurtled towards Earth when two other asteroids collided. Those asteroids themselves were left over from the vast accumulations of dust and debris that formed our own solar system, making the Wold Cottage a scrap of a scrap from the days when our Sun was a new star.

Moon rocks

If the Apollo 11 mission is famous for anything, it's Neil Armstrong's proclamation "One small step for man, one giant leap for mankind". But the mission to our lone moon wasn't simply about planting a flag in the lunar dust. With boots on the Moon, the Apollo 11 mission was able to gather samples of the Moon's rock that have helped scientists better understand the formation of our closest neighbour in space.

When the Apollo 11 astronauts returned to Earth in 1969, about 22 kilograms of soil, rocks and core samples came with them. Nobody had brought stones back from space before, and experts were a little worried they might carry some unknown microbe or pathogen that would be dangerous to life here on Earth. Nobody wanted a *War of the Worlds*-type conclusion for our own species. Lab mice were injected with samples of lunar rock, cockroaches were fed lunar dust, and other tests were carried out to make sure that it was safe to keep the lunar samples on Earth. Once the samples were cleared for study, they provided researchers with crucial evidence of what the Moon was made of – hint: not cheese – and how it formed. This is how the discipline of selenology, or lunar science, got its start.

Landing site

The Moon has many geologic features. There are impact craters, domes, grabens, rilles and more, creating a complex surface. But much of what we know about the Moon's geology comes from flatter areas. Some of the stones collected by the Apollo 11 crew came right from their doorstep at their landing site in the Mare Tranquillitatis, or Sea of Tranquility. This 873-kilometre-wide dark area was given its aquatic-sounding name centuries ago, when early astronomers thought it might be an ocean. But that glossy

surface was rock, smooth enough to make it an attractive landing site for the Apollo 11 mission.

Researchers who studied these samples soon realized that the oceanic appearance of the Mare Tranquillitatis came from the type of rocks found there. The rocks in this area are primarily made of basalts – porous rocks that form when molten rock rapidly cools. Such rocks are often the product of volcanic eruptions on Earth. Whatever happened in this lunar region, at least, involved an incredible amount of heat.

But the basalts weren't all. The astronauts also recovered stones called breccias – rocks made of smaller fragments that have been later cemented together. These rocks had a more violent history than many of their earthbound counterparts. As you might guess from its many craters, the Moon has been repeatedly bombarded by various chunks of space rock and debris over time. Those impacts shattered stone and then fused the shards together into breccias, which were a mix of what was present where the meteorite hit.

What researchers found in those breccias added another layer to the Moon's history. Some of the fragments in the breccia contained a rock called anorthosite, a kind of basalt. More specifically, the

Opposite A photograph of the relatively flat surface of the Mare Tranquillitatis taken during the Apollo 11 mission.

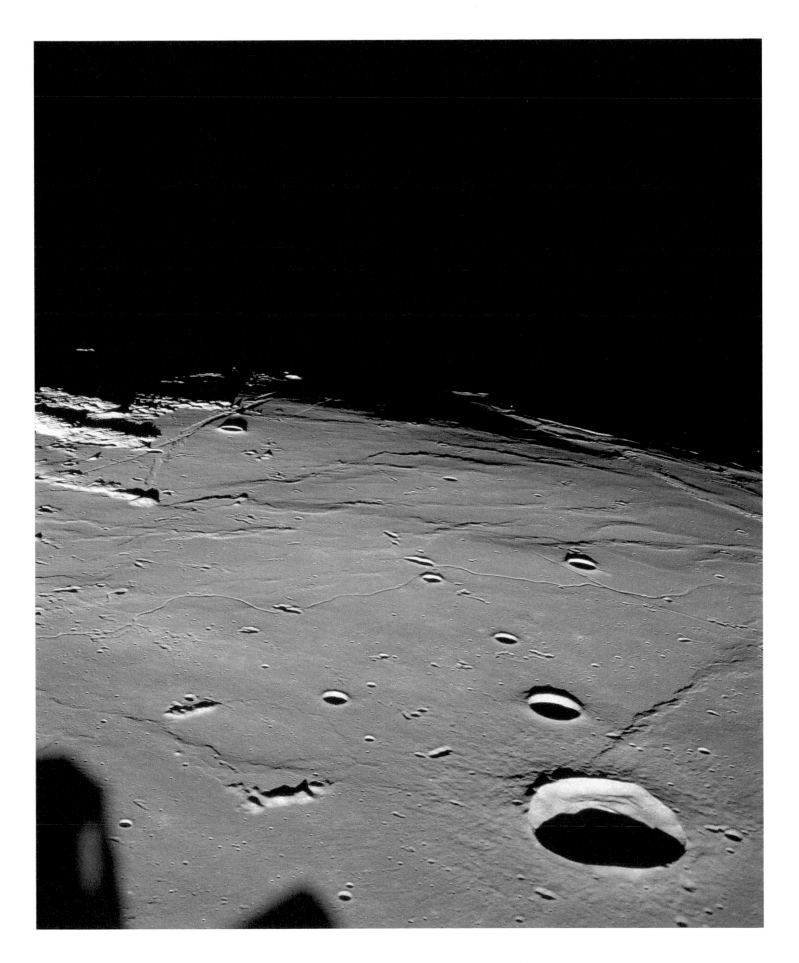

make up of the rocks hearkens back to a time when the Moon's surface was molten – a vast magma ocean of melted rock. The state of the Moon was so viscous, in fact, that Earth's gravity altered its shape. Around 4.5 billion years ago, the young Moon had a more oblong shape. Only after the Moon began to cool did it take its more circular shape that we see in the sky.

Heavy bombardment

The Moon's geologic trauma was hardly over, though. The stones recovered from the lunar surface indicated that the Moon was repeatedly smacked with other rocks and space debris through time. The Mare Tranquillitatis itself was outlined by such an impact. An immense chunk of rock – a large comet or asteroid – struck this part of the Moon about 3.9 billion years ago, melting the rock and creating the geological profile of the area today. Selenologists refer to this time as the lunar cataclysm, or late heavy bombardment.

The emerging picture is that the Moon started off as a ball of molten rock that cooled over time, rock layers eventually separating out and chilling to become the round, reflective body in orbit around our planet. While dates from one sample to another vary somewhat, they are all around 4.35 billion years old, dating from 200 million years after the formation of the Earth – a hint that the Moon went from liquid stone to cold, orbiting rock very quickly.

Above Lunar Sample 10021,79, a small rock from the very first sample collected from the Moon during the Apollo 11 mission in 1969.

Left Astronaut Gene Cernan collects rock samples during the Apollo 17 mission, the sixth and final Apollo landing in 1972.

Zircons

Even though you can't hear it, the clock inside zircon crystals is constantly ticking away. These tough mineral formations help geologists to date rocks to a high precision, forming part of a many-pronged strategy.

Geologists use many different methods to determine a rock's age. Some of these techniques date back to the seventeenth century, when scholars and naturalists began to understand a concept now called the law of superposition.

Picture a stack of grey, maroon and purple rocks exposed in a big, towering mesa at the edge of a desert. Each differently coloured section is a separate layer, and the rocks were all laid down a very long time ago as part of an ancient floodplain where river water deposited sand and mud. According to the law of superposition, a geologist would hypothesize that the rocks at the bottom of the stack would be older than those on top. Back when the rocks formed, new sediment in this area was laid down on top of the old, like putting one sheet of paper on top of another until you have a great big stack. In most cases, rocks at the bottom of such a stack are older and those near the top are younger.

There are exceptions, of course. Sometimes stacks of rocks become titled or flipped long after they have formed under pressure from mountains pushing up or faults that run through the Earth's crust. For this reason, experts often turn to other lines of evidence to understand how long ago particular rocks formed.

Clues from fossils

Sometimes fossils can help. In a discipline called biostratigraphy, a researcher can look at all the different species found in a rock layer and see if those species are found in another rock layer elsewhere.

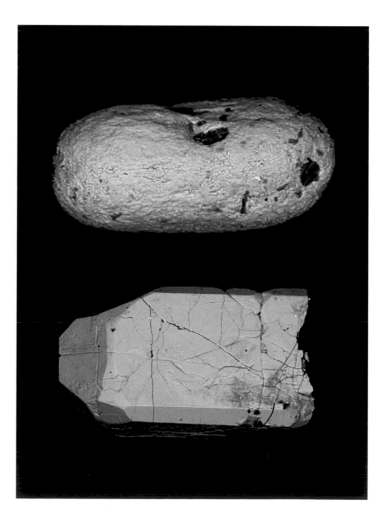

Let's say that a paleontologist finds a species of coil-shelled squid relative called an ammonite in a rock layer about a metre above a valley floor. If that researcher finds that same species of ammonite in a different rock layer many kilometres away and higher up in the rock section, they can be confident that those two rock layers represent the same time period even if they're far apart.

Both of these approaches – the law of superposition and biostratigraphy – are good for what's called relative dating, or figuring out whether a rock layer is generally older or younger than others in a stack. But if we want to work out how many millions of years ago a rock formed, what experts call the absolute age, we need to zoom in on clues hidden within the rock itself.

Resilient crystals

Tiny crystals called zircons are the key to determining absolute age. These microscopic crystals often form in magma within the Earth. They are carried to the surface by venting and eruptions. And they're tough. Zircons are resistant to breakdown and persist inside rocks that contain prehistoric lava or ash. When paleontologists dig

up a bonebed of dinosaurs buried in the ash of an ancient volcano, for example, they're excavating rocks full of these helpful crystals.

It is their deep-Earth origin that makes zircons so important. From their molten rock beginnings, zircon crystals contain uranium, a radioactive element. The neat thing about uranium is that it changes into lead at a very slow, constant rate. Geologists measure this rate as uranium's half-life, which is the amount of time it would take a given amount of uranium to halve itself as it decays into lead. By looking at how much uranium a zircon crystal contains versus how much lead, you can calculate just how long a rock has been around and obtain a more precise date for when it formed. The fact that uranium has a half-life of 4.5 billion years has made it very useful for dating some very old rocks.

Above Ammonite fossils are distributed widely around the world, making them particularly useful to biostratigraphy.

Opposite A backscatter electron micrograph of two zircon crystals from the Jack Hills.

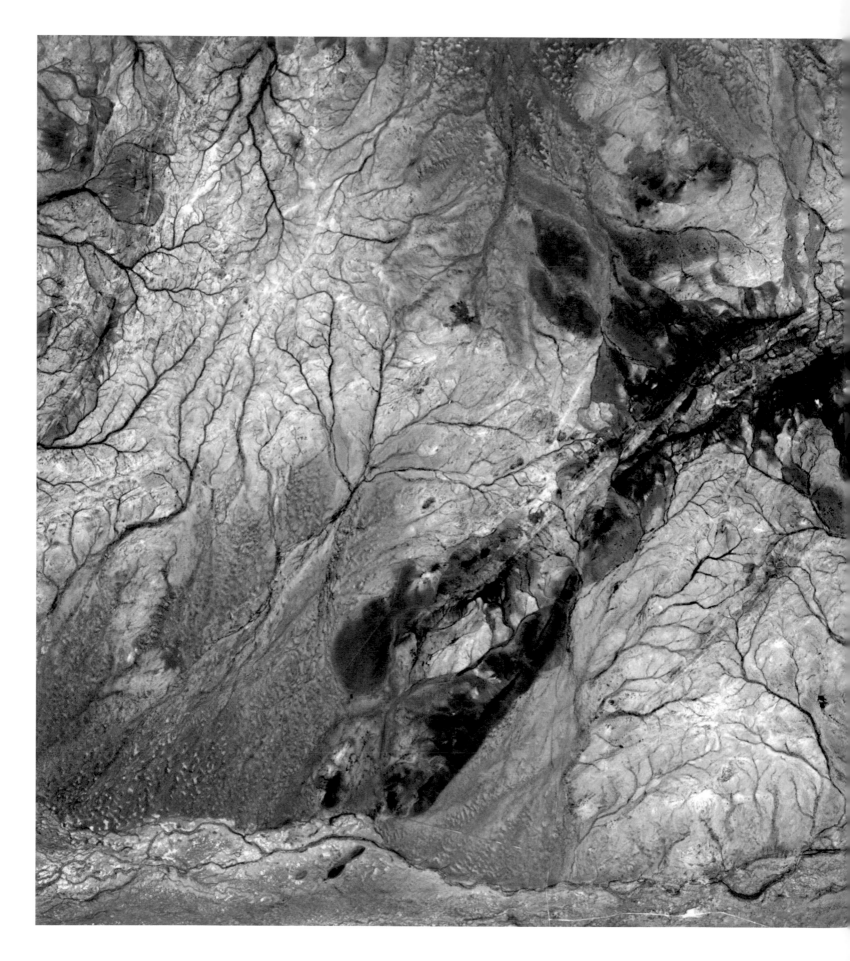

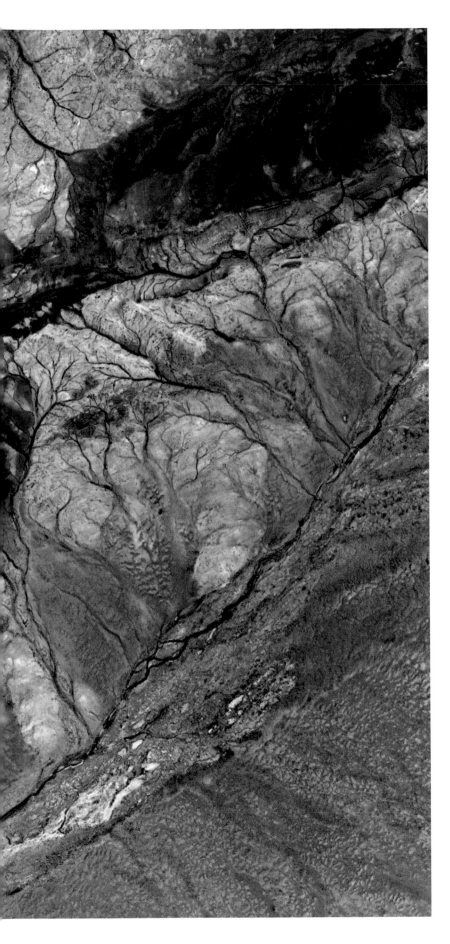

Combining evidence

Not all rocks can be dated from zircon crystals. Rocks made out of Jurassic mud from an ancient seabed may not have the proper components to date. But there have been enough eruptions throughout Earth's history that many layers of sedimentary rock are sandwiched between, or in close connection to, the kinds of rock that contain zircon crystals. Date those rocks and you can start putting boundaries on the time frame, perhaps even using fossils from a deposit in one place to examine another with the same organisms that happens to have zircon-bearing rocks, in order to come up with a more exact date.

This scientific detective work has allowed geologists to identify some of the oldest rocks on Earth. Testing zircons has enabled researchers to identify strata from the Jack Hills in Western Australia as the oldest known on Earth — over 4.3 billion years old. Given that the Earth itself is estimated to be just under 4.6 billion years old, that's awfully close to the beginnings of our planet.

Above A quartz-pebble metaconglomerate from the Jack Hills. This rock is nearly 4.3 billion years old.

Left A satellite image of the Jack Hills in Western Australia.

Acasta Gneiss

Geology does not tell a neat, ordered story. The Earth isn't like an onion, made up of even layers stacked from oldest to youngest. Rather, the crust of our planet is a jumble in which uplift, erosion and other forces have buried some parts of ancient history and brought others to the surface. These constantly working forces have exposed incredibly old rocks in the vicinity of Yellowknife, Canada.

These great blocks of dark, striped rock are called the Acasta Gneiss Complex. Cut open and cleaned up, the black, grey and tan colours seem to create swirls through the stone. The minute details have allowed geologists to determine just what type of rock this is and how it formed during the very early days of the Earth.

When geologists classify a rock, they often lump it into one of several broad categories based upon how rocks form. Igneous rocks are once-molten rock. When lava from a volcanic eruption cools and solidifies, it becomes an igneous rock. Sedimentary rocks are very different. These are rocks like sandstones and mudstones that form when lots of small particles pile up and create layers as they solidify over time. (These are the sorts of rocks to look in if you're searching for fossils.) Metamorphic rocks are like a mix of the two. Metamorphic rocks start as igneous or sedimentary rocks that are then further altered by extreme heat and pressure underground. For example, sand that piles up on a seafloor will become sedimentary rock, and if those rock layers are shoved beneath the surface to higher temperatures and pressures, they will transform into metamorphic rocks.

Gneiss is a metamorphic rock. The ancient rocks found outside Yellowknife originally formed as something else before being transformed. Despite all that pressure and time, though, the Acasta

Gneiss still preserves some zircon crystals that are capable of providing a date. It turns out that the Acasta Gneiss dates back to about 4 billion years ago, within about 600 million years of when the Earth itself formed.

Above A 4 billion-year-old Acasta Gneiss.

Opposite A satellite image of Slave Crayton in Canada's Northwest Territories, which contains the Acasta Gneiss.

The young Earth

Geologists call this time the Hadean Eon. The word is derived from Hades, the Greek god of the underworld, and tells you something about what our planet was like back then. It was an age when the Earth was very hot and still settling down from its formation. Experts estimate that the early planet's surface temperature was 230 degrees Celsius, but somehow early oceans spread across the surface despite the great heat. The Earth was slowly transitioning from a static planet with a hot molten surface to a cooler one on which plate tectonics shuffled around the great continental crusts.

The way our planet works is that the rock crust is really just a thin rind on the outside of deeper and increasingly hot layers. The continental crust is made up of many separate plates on top of a constantly moving mantle that, despite being mostly solid rock, has the consistency of caramel. These characteristics mean that the plates move, bump each other, slide beneath one another, and are pushed by places where the seafloor is spreading out thanks to the addition of new molten rock from within the Earth. Plate tectonics is the reason continents such as Africa and South America were once connected but no longer are.

Plate tectonics was still a relatively new phenomenon on Earth when the Acasta Gneiss formed. The metamorphic nature of these rocks is a clue. Researchers hypothesize that the Acasta Gneiss indicates that the Earth was recycling its crust.

From the Earth's early molten state, the rocks of its crust began to cool. The lower, goopy mantle layer outside the core allowed the plates of crust to move. And at some places where plates meet, one plate is sometimes pushed beneath the other. This puts the weight of the upper crust on that layer and subjects it to more heat, transforming the rock. After the original Acasta rock layers formed, geologists hypothesize, the surface layers of rock were slowly ground down into the Earth's mantle before that melted and altered stone was brought back out again, like a giant geological conveyor belt. The Acasta Gneiss tells part of that story. Researchers believe that the rocks come from granite, which once formed early continents and was later altered by pressure and time. What was once drowned by stone was eventually brought back up to the surface, bringing with it a record from the nascent Earth.

Above A coloured scanning electron micrograph of the Acasta Gneiss.

Opposite A slab of gneiss displaying the dark bands that are typical of this metamorphic rock.

Impossible life forms

Deep in the ocean, where light doesn't penetrate, there are echoes of what the Earth's earliest ecosystems might have been like, feeding on energy that is constantly seeping out from the planet itself.

In 1977, oceanographers aboard the deep-sea submersible *Alvin* were looking for hydrothermal vents around the Galápagos Islands in the Pacific Ocean. These vents are places where magma is close to the seafloor, heating water that comes up through fissures and pipe-like chimneys. When the underwater explorers finally saw one, they were astonished. Despite the fact that no sunlight reached them, the vents were teeming with life. There were ghostly-white crabs, strange fish, 3-metre-long tube worms with rosy-red tips, and vast mats of bacteria – an entire community of organisms no one ever expected, and ones that might hold the secret to early life on our planet.

On most of the planet, plants form the basis for complex ecosystems. Plants photosynthesize – vegetation absorbs sunlight and uses that energy to create its own food from carbon dioxide and water. Other organisms feed on the photosynthesizers. But in the oceans, sunlight doesn't reach below 200 metres. Photosynthesizers can't survive there. The vent organisms had to rely on another source of energy called chemosynthesis.

Below Anomuran crabs mass around a hydrothermal vent at a depth of 2,400 metres.

Opposite A gassy 'white smoker' vent in which the temperature of the water is more than 100°C.

Bacteria are the key to these odd environments. Bacteria – either out in the ecosystem or within the bodies of animals – are able to use the energy stored within the chemical bonds of compounds coming from the vents, such as methane, to create their food. This means that the strange creatures found around hydrothermal vents, like those immense tube worms, aren't catching food to eat so much as hosting bacteria inside their bodies that are able to provide food. It's a totally different way of life, and this chemosynthetic process might be a clue as to what Earth's earliest life was like.

Microscopic fossils

Ever since the discovery of the deep hydrothermal vents, researchers have wondered whether life on Earth might have started the same way. Fossils found in Quebec, Canada, dating from more than 4 billion years ago, lend the idea some weight. Inside reddish crystals rich in iron, scientists have found tiny tubes that bear a striking resemblance to structures created by microorganisms that live around hydrothermal vents today.

When it comes to very old fossils, experts are often divided over their interpretations. Many tiny fossil structures, such as tubes or rounded structures that look like cells, can be created by other natural processes and are not signs of life. For this reason, multiple lines of evidence are needed to confirm the identity of the world's oldest fossils. Experts investigate the shapes of these structures, their chemical signatures and other details to ascertain whether there's a match between the fossils and forms of life that have previously been seen. In this case, the experts were not looking for animals or any form of complex multicellular life, but for matches between the patterns in the crystals and the patterns that microscopic, single-celled organisms are capable of creating.

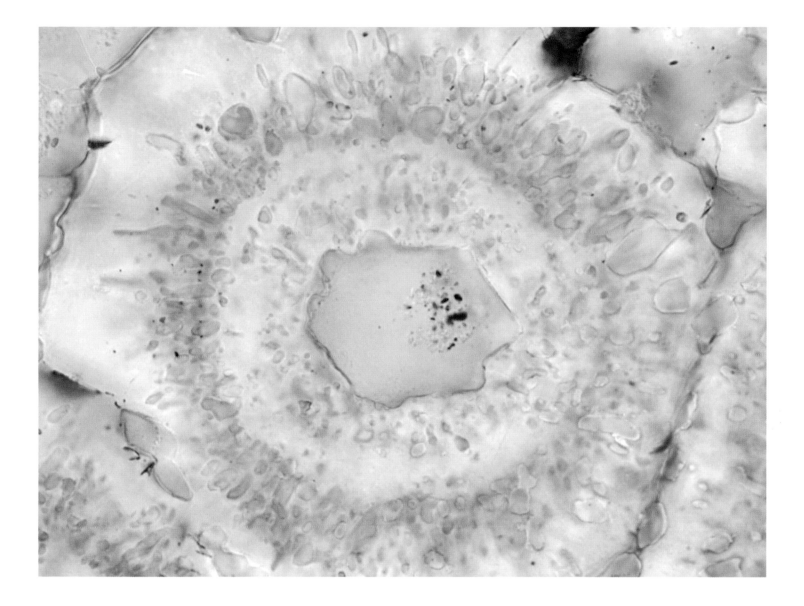

The first life on Earth?

The rocks in which the fossils were found were collected from the shore of Canada's Hudson Bay. They were formed when a prehistoric vent oozed lava over the seafloor. It's a similar kind of environment to the deep-sea vents that gush and fume today, providing the essential methane and sulphur-rich compounds that chemosynthetic bacteria rely on to make their food. Inside the rock containing the tubes are carbon compounds, phosphorus and oxidized iron, all with a chemical signature that suggests these elements were left there by living processes rather than geological happenstance.

Experts are split on their interpretations of these rocks, and any find relating to early life is sure to attract plenty of scrutiny and repeated testing. But if these crystalline clues are real signs of life, then they may go a long way towards filling in the context for some of Earth's earliest life forms. Life may not have started in a shallow, sunlit pool, but deep in the ocean where the heat of the inner Earth broke through the surface and provided the raw materials for primitive life forms. A century ago, no one had any idea that such organisms could exist now, much less in the deep past. The vent bacteria are another sign of life's resilience and resourcefulness, and a clue that the origins of life on our planet may be much stranger than anyone ever anticipated.

Above Rocks containing possible fossils of microbes that lived around ancient vents.

Opposite A colony of *Riftia* tube worms around a vent near the Galápagos Islands.

Stromatolites

If you want to travel back in time to get a peek at what the Earth was like billions of years ago, there are a few places where you can do just that. Lagoa Salgada in Brazil, the Great Salt Lake in the western United States, Lake Salda in Turkey and Shark Bay in Australia can all offer you a glimpse. These places are home to a peculiar kind of ancient life called stromatolites.

Seen from the outside, stromatolites might look like rocky pillows or towers covered in a kind of fuzz – fittingly, the word stromatolite roughly translates as 'mattress rock'. Planted in place, they are not exactly the most dynamic of structures, but they are far from inert. These bulbous mounds of stacked stones were created by cyanobacteria, a form of bacteria capable of photosynthesis and one of the oldest components of life on Earth.

We are lucky to be able to see and study modern stromatolites, and geologists have a pretty good idea of how these structures form. The way stromatolites form today is much the same as it was billions of years ago. As they go about their biological business creating food from sunlight, carbon dioxide and water, clusters of cyanobacteria make a natural adhesive. This goopy output does two things. The glue not only helps keep the cyanobacteria together, but it also traps sediment similar to the way the roots of a plant trap soil. And, in fact, sometimes the cyanobacteria precipitate minerals of their own.

All this activity makes a thick microbial mat on top of stuck-together sediment. And the cyanobacteria don't just stay in one spot. They're drawn to the light, essential for making food, so they move upwards to stay on top of the sediment accumulating below. Over time, the older layers will start to turn to stone, creating a raised platform with the living cyanobacteria on top. The whole structure resembles geological strata, with the oldest layers below and the newer ones directly beneath the cyanobacteria.

The conditions for life

Stromatolites are found at fossil sites around the world, but the oldest ones yet uncovered are from Western Australia. These mushroom-shaped towers date to about 3.5 billion years ago, in the Achaean Eon, the second geologic eon that followed the fiery Hadean Eon and lasted from about 4 billion years ago to 2.5 billion years ago.

By the Archaean Eon, the Earth's crust had cooled sufficiently for continents to form. The landmasses were not in their present positions, but large blobs of rocky land dotting a vast global ocean. The atmosphere wasn't anything we'd consider comfortable. Methane fell from the sky in big wet drops, creating a haze all around the planet. Even the light was different. The Sun was about 70 per cent as bright as it is now, but greenhouse gases trapped the Sun's heat to prevent the world from freezing over. All the same, there was enough light for life to start doing something new. At the time, photosynthesis was still new. Before this time, life didn't need oxygen but instead made food through chemosynthetic pathways. That microorganisms came to use sunlight to help them

make the sugars they relied on for food was a chance event that happened to take off, one of evolution's successful experiments.

Given what biologists and geologists have come to understand about modern stromatolites, researchers expect that these blobs of sun-hungry cyanobacteria were very important to shifts in the ancient atmosphere. Oxygen is one of the products of photosynthesis, and as stromatolites proliferated through billions of years, they helped introduce greater quantities of this element, crucial to the process of respiration. We wouldn't know much about these changes were it not for the stromatolites that cyanobacteria created, structures categorized as microbialites by geologists since they are rocks shaped by the activity of microbes.

That stromatolites have hung on for so long is a wonder. The structures hit their high point 1.25 billion years ago but had vastly scaled back by about 550 million years ago, when animal life began to rapidly evolve into many new forms. The world stromatolites helped to create may eventually have been their undoing. An increasingly oxygenated Earth helped set the right conditions for grazing creatures like early snails and other invertebrates, which found the fuzzy mats of cyanobacteria atop stromatolites to be a tasty buffet. Stromatolites still hold on in a few places, especially extreme and isolated spots – a whisper of the ancient Archaean world.

Top A 3.4 billion-year-old fossilized stromatolite from Pilbara Craton, Western Australia.

Above A close-up of a fossilized stromatolite.

Above Stromatolites in Shark Bay, Western Australia.

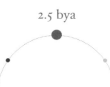

Hintze Hall Pilbara iron rock

It's easy to take oxygen for granted. The element is in every breath we take, forming 21 per cent of our atmosphere. But this wasn't always so. For billions of years, Earth's atmosphere was relatively oxygen-poor. The great change, occurring about 2.5 billion years ago, came from life itself. A huge piece of banded rock in Hintze Hall in London's Natural History Museum is a testament to this turning point.

While a piece of banded rock full of iron might not seem like the most vivacious of objects, the stone is a clue to what life was doing at the start of the Paleoproterozoic Era (2.5 billion to 1.6 billion years ago). Life had already existed on Earth for more than a billion years at this point, with the mattress-like stromatolites becoming more prominent in shallow waters around the planet. But it was at the 2.5-billion-year mark that more forms of single-celled organisms began to photosynthesize and produce oxygen. Great blooms of cyanobacteria and other photosynthesizing microbes suffused the seas with it. As those organisms turned sunlight into food for themselves, the oxygen they produced combined with iron kept in solution in seawater.

The key was a process called oxygenic photosynthesis. It works like this. When photons from the sun meet chlorophyll inside cyanobacteria, the sunlight provides enough energy for the cyanobacteria to take some electrons from water to create both hydrogen and oxygen. The electrons are used by the cyanobacteria to create a compound called adenosine triphosphate, or ATP, which can transport energy through the cell. From a global perspective, incalculable numbers of cells undergoing this process every day started to create a great deal of oxygen.

Running out of iron

During the Paleoproterozoic, the world's oceans were relatively rich in dissolved iron. In solution, this is called ferrous iron. But when iron is oxidized, with oxygen nabbing another electron from the iron, the metal changes. It becomes ferric iron, which then precipitates out and sinks. As the oxygen joined with the iron, the new molecules of iron oxide sank to the seafloor in layer after layer through the years. Just like stromatolites created layers beneath them, photosynthesizing bacteria created banded layers of iron rock until the majority of the iron in the water was used up. At that

Left The Pilbara rock on display in Hintze Hall in London's Natural History Museum.

Opposite Layers of iron ore in Hammersley Gorge, Pilbara, Australia.

point, the oxygen was free to do something else. It went up, into the atmosphere, creating what researchers have called the Great Oxygenation Event.

The chunk of banded iron in Hintze Hall, excavated from an exposure in Pilbara, Australia, dates from the Great Oxygenation Event. In boulders like this, the darker layers are often made of an iron mineral called magnetite, while the red ones are a form of silica called chalcedony, the part that gets its blushing shade from the buried iron oxide. The alternations between the two – dark and red, dark and red – are probably attributable to changing seasons on the ancient Earth. During warmer, sunnier months, more red bands formed. During colder and darker months, the darker bands formed. This is another clue that this immense rock was partly created by the activity of early life. The irregularities likely represent times when photosynthesizing bacteria proliferated and then died back, creating a mosaic of growth patterns left behind in the stone.

The rock from Pilbara is far from the only one of its kind. Banded-iron formations have been found on every continent. This was a worldwide event that changed Earth forever. And the results were not wholly positive for life.

Changing the course of life

The Great Oxygenation Event is sometimes referred to by other, less flattering names, such as the Great Oxygen Catastrophe, because, at the time, oxygen was deadly to many forms of early life other than cyanobacteria. Prior to the spread of cyanobacteria most life on Earth was anaerobic, meaning that it did not require oxygen. That was useful during a time when oxygen was rare or bound with abundant iron. Once oxygen began to escape into the atmosphere, bringing the world's oxygenation close to today's marker of 21 per cent, the anaerobes found themselves at a distinct disadvantage.

The photosynthesizing organisms effectively created a poison that killed off many other single-celled organisms, and almost led to their own undoing. The new glut of oxygen in the atmosphere combined with methane to create carbon dioxide. While both are greenhouse gases, methane is a more effective one than CO_2, meaning that the atmospheric shifts caused the Earth's temperatures to swiftly plummet. The organisms that couldn't handle oxygen were relegated to oxygen-poor, deep-sea environments, while single-celled species capable of using oxygen proliferated. The banded iron left behind represents a boom followed by a terrible bust, one that forever altered which forms of life could thrive on Earth.

Left Layered rock in Joffre Gorge, Pilbara, Australia.

The Grand Canyon

Running through the high desert of northern Arizona, this immense geological formation has names in Hopi, Navajo, Yavapai and Spanish. To most, however, it's known as the Grand Canyon – a 446-kilometre-long gash in the rock that plunges down more than 1,800 metres at its deepest parts.

While certainly famous as an often-crowded tourist spot, the Grand Canyon is also a geologist's dream. Much of what we understand about the Earth and its history comes from rocks that have been fortuitously exposed in places where we can get to them. It's difficult to understand the backstory of a place when it's covered by a lawn or car park. But here, erosion has sliced through the rock like a huge, stony layer cake to reveal snapshots of Deep Time.

Layers of time

The canyon isn't a static monument to ancient ages. Like other geologic wonders in the desert, it was carved out by water and time. Around five million years ago, the ancient Colorado River flowed over this area. Just as it does today, the river carved out stone beneath it, sinking ever deeper and becoming narrower as it cut down through the stone. Millions of years of erosion carved the canyon out of the ancient rock layers, revealing hundreds of millions of years of history stacked on top of each other. Take a hike from the canyon rim to the bottom and you are travelling through time.

The top of the Grand Canyon is already pretty old. When you're on the overlooks, gazing at the rocky wilderness beyond, you're standing on the 280 million-year-old stone of the Kaibab Formation. These rocks formed during the Permian Period, a time before dinosaurs when our protomammal ancestors and relatives were the dominant animals on land. Paleontologists are still making discoveries in these layers. In 2019, experts announced that they had found the tracks of a lizard-like animal called a diadectomorph, which had ambled through this area back when North America was part of the supercontinent Pangaea.

Left A sample of folded Vishnu basement rock on display in the Grand Canyon National Park.

Opposite The layers of rock take you back in time to 1.8 billion years ago as you move down the cliffs.

Dip down through the canyon and you go further and further back in time through the Toroweap Formation, the Coconino Sandstone, the Supai Group and more, all the way to the bottom. This is where the oldest rocks of the canyon are found – igneous rocks formed during the Precambrian about 1.8 billion years ago.

Clashing landmasses

Many of the Grand Canyon's rock layers are sedimentary, formed by the gradual accumulation of sediment in environments like streams and lakes. But the lowest rocks had a much different origin. Experts call these foundational stones the Vishnu Basement Rocks, named after the Hindu deity who created the universe. The origin of these rocks was no peaceful event, however. Over 1.8 billion years ago, the area where the Grand Canyon now sits was part of an ocean basin near an area where prehistoric mountains were beginning to be pushed up. Ocean sand, ash from volcanoes, silt washed in from rivers and other forms of sediment all mixed together here, and they were set to undergo a great change.

Though the collision of landmasses happens very, very slowly, at rate of centimetres a year, it still feels like a violent interaction. About 1.8 billion years ago, at least two different chains of islands that had been created by volcanic activity – like the Hawaiian Islands today – smashed into prehistoric North America. Ocean sediments were lifted out onto land, only to be subsumed by later tectonic movement and buried almost 20 kilometres beneath the surface. The heat from the Earth's mantle and the pressure from being squeezed beneath the crust changed those ocean sediments into metamorphic rocks, which geologists now know as the Granite Gorge Metamorphic Suite.

Ongoing processes

These rocks weren't done changing yet. The damage of the landmass collisions, with volcanic islands running into ancient continents, caused magma to ooze out of the places where one part of the crust was being driven under another. Geologists know this as a subduction zone. In this case the magma seeped out of the Earth and into the already-formed rocks to form igneous intrusions into these rocks. The story isn't one of orderly accumulation, but the push and pull of our planet grinding itself down, ancient rocks transforming over immense timescales to be exposed much later by the action of rushing water. And the process is ongoing. What will eventually become of these storied rocks is yet to be told.

Above The Horseshoe Bend on the Colorado River in the Grand Canyon.

Mitochondria

Some of the starkest evidence of past ages comes in forms like towering stacks of ancient rocks or the burn of distant stars. But there are also signs of Deep Time within our own bodies. Inside our cells, tucked away with the other essential organelles, is a little sausage-shaped body called the mitochondrion, which has a fascinating history dating back nearly 1.5 billion years.

During our day-to-day lives, mitochondria create the chemical energy we need to run our essential biological functions. Chemicals inside the body are converted by the mitochondria in our cells into ATP, or adenosine triphosphate – the molecule that provides energy for other cellular reactions. For this reason, the mitochondrion is characterized as 'the powerhouse of the cell'.

Double membrane

Aside from their essential function, there is also something strange about mitochondria. They are surrounded by not one, but two, membranes. No other organelle in our cells is like this. It turns out that this double wrapping is a sign of an ancient and momentous event that has only ever happened once in the history of life.

For most of the history of life on Earth, there were no complex, multicellular life forms. Most early life prior to 650 million years ago – the vast majority of life's history – consisted of microscopic, unicellular organisms that either made their own food or consumed food from their environment. Cyanobacteria are an example of producers – cells capable of making their own food through photosynthesis – while cells that eat other cells are categorized as consumers, or heterotrophs.

Endosymbiosis

One way heterotrophs caught and consumed food was by a process called phagocytosis. Think of a single-celled organism like an amoeba. This organism could open a pocket or fold that surrounded that little morsel of food, totally surrounding it and bringing it into the body for digestion. Then at some point in history, around 1.45 billion years ago, one of these predatory cells tried to engulf another cell, only to have something entirely unexpected occur. The would-be dinner was taken into the cell and stayed there to become a part of it – a phenomenon known as endosymbiosis.

Instead of providing a quick meal, mitochondria started to produce energy for the host cell. This is known as the endosymbiotic theory. As these ancient cells reproduced, the mitochondria went along for the ride. They even have their own form of DNA, called mitochondrial DNA, which is transmitted along with nuclear DNA when cells reproduce.

Opposite Top Mitochondria inside a mammalian lung cell.

Opposite Bottom This false-colour transmission electron microscope micrograph shows the many mitochondria (blue) in the cytoplasm of a hapatocyte, the main type of liver cell.

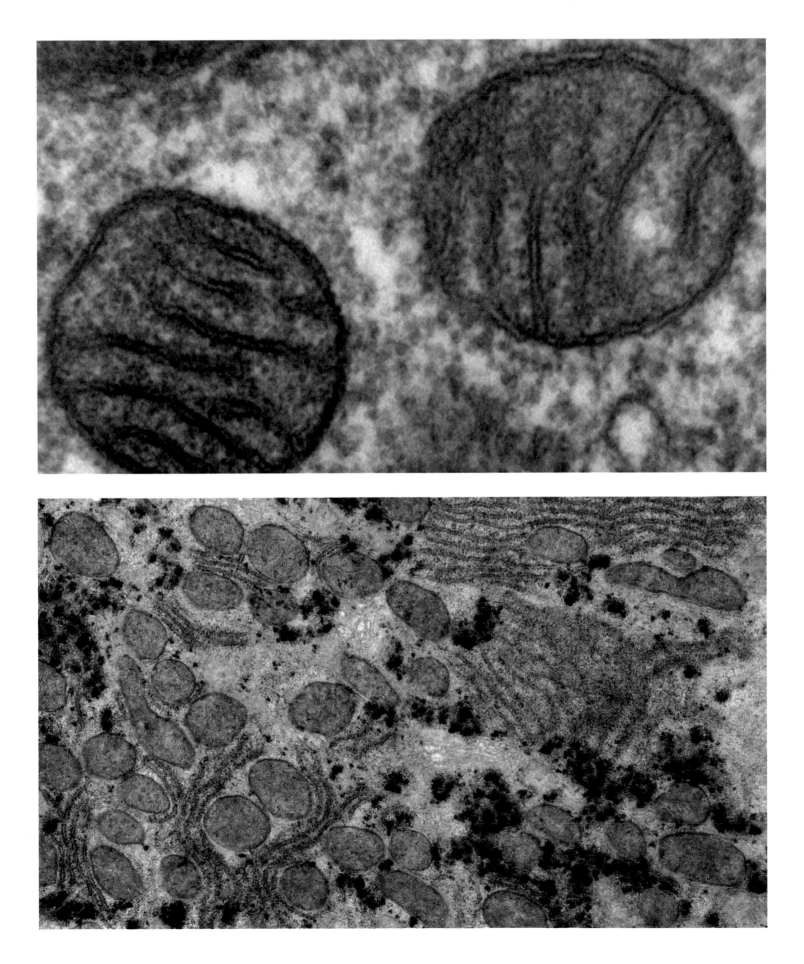

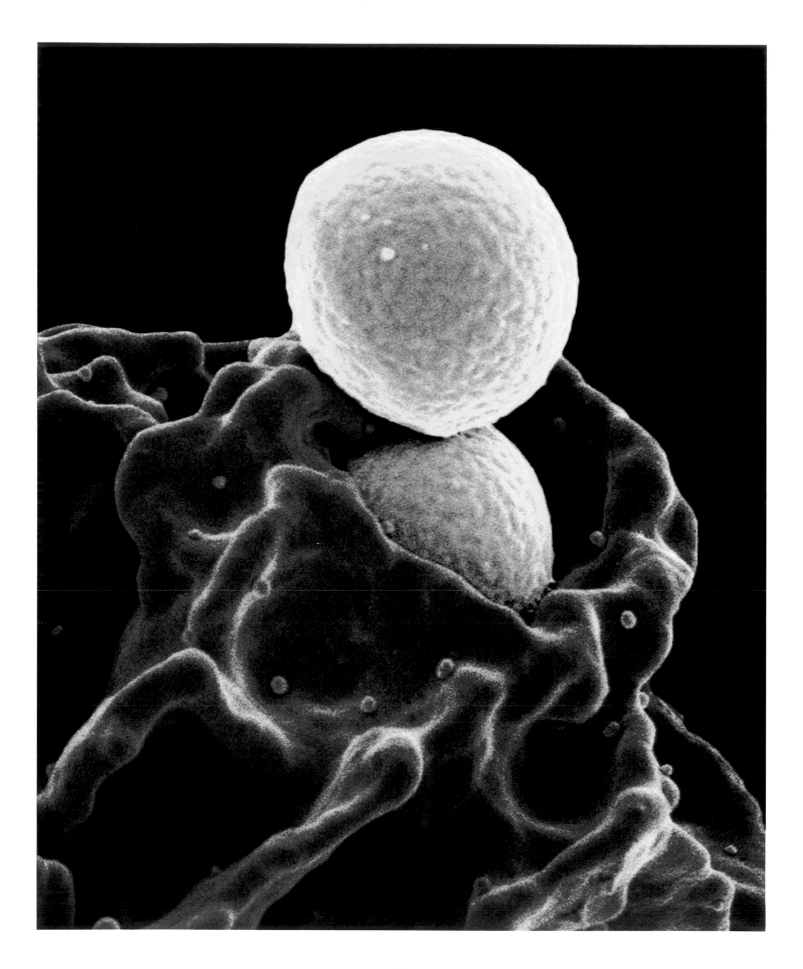

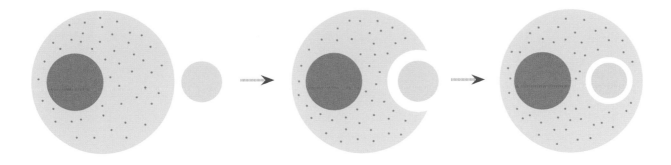

That mitochondrial DNA allowed the organelle to make a permanent home inside host cells. However, the single instance of failed digestion wasn't enough. The host cell had to be able to replicate the mitochondrion and provide the essential parts the organelle needs to work. What probably occurred, biologists hypothesize, is that some of the mitochondrion's DNA was transferred into that of the cell nucleus. Not only did the physical body of the mitochondrion become part of the cell, but now the cell had the genetic instructions to make mitochondria as part of its ongoing generations.

So far as researchers have been able to discern, mitochondria reside only within eukaryotic cells. These are cells that have a clearly encapsulated nucleus that contains the bulk of the cell's DNA. (Prokaryotes, the other major cell type that includes bacteria, don't have nuclei or organelles surrounded by membranes.) The number of mitochondria within cells can vary greatly, even in the same organism. In our own bodies, for example, blood cells don't contain any mitochondria while our liver cells contain more than 2,000 mitochondria each.

Much of what we know about the mitochondrion and its history involves studying how the organelles work today. Finding fossils of these ancient cells is difficult work. Despite the fact that single-celled organisms were everywhere – and continue to be! – conditions had to be just right to preserve them in rock. By looking to genetic evidence, though, researchers have estimated that the origin of the mitochondria in our cells dates back to about 1.45 billion years ago.

Above The three stages of phagocytosis, by which one cell ingests another.

Left American biologist Lynn Margulis (1938–2011) first proposed the theory of mitochondrial endosymbiosis in 1967. Her idea only gained acceptance many years later.

Opposite White blood cells destroy bacteria such as MRSA through phagocytosis.

Single evolution

If such an event happened once, could it have happened again? Evolution thrives on variation, after all, and it's strange to think that only one cell had just the right conditions for mitochondria to be welcomed into the internal goop of the cell. But that's probably what happened. There is no evidence that there were multiple forms of mitochondria or any sort of experimental phase during which varied cells brought the ancestral mitochondria inside.

The best evidence for this singular origin comes from DNA. Not only do mitochondria have their own DNA, but all known eukaryotic cells share common mitochondrial genes. This is not a case of convergent evolution, but particular genes that mitochondria share whether they're in a pine tree or a *Euglena* whipping around a petri dish.

William Smith's Map of England

Any geologist worth their weight in igneous rocks knows the name William 'Strata' Smith. The nineteenth-century geologist is a legend in the field, best known for a historical first. In 1815, Smith unveiled the first geological map of an entire nation – Smith's Map of England.

Geological maps seem commonplace today, as close as a gas station map rack or a few coordinates punched into Google Earth. Two centuries ago, however, working out which rocks lay beneath a particular area required a great deal of dedicated work. Early geologists had to 'walk the outcrop', as they say, seeing which rocks were exposed and where – but things became tricky in places where rocks dived beneath areas covered by fields, towns or other obstructions, or where the same layer seemed to be separated by gaps.

Fossil clues

William Smith had a solution to these problems. Many rock layers in England are fossiliferous – from the Jurassic beds that produced icons like *Megalosaurus* to Ice Age deposits that contain the remains of hyenas and elephants that once called the island home. More specifically, many of England's geological strata host fossil invertebrates that are crucial for the science of biostratigraphy.

Working over a century before absolute dating techniques using zircon crystals would be invented, Smith's ability to correlate rocks with each other through space and time rested upon the fossil organisms found within those layers. He picked up the idea from his time digging ditches. During a period working on canals in England, Smith noticed that fossils appeared in a particular order through rock layers. If you knew their order, you could figure out whether you were looking at an older or younger rock by the fossil inside it. Smith called it 'the principle of faunal succession', and this principle would eventually lay the foundation of his epic map.

In time, Smith personally mapped an area over 175,000 square kilometres, encompassing the whole of England and Wales. The resulting image, with rock layers highlighted in 23 different tints, was the world's first comprehensive geological map of a country – one that stretches 2.6 metres tall and 1.8 metres wide. (And that's not counting extras like detail of the rock layers between London and Snowdon, showing how the rocks of southern England dip to the southeast.) The full title Smith gave his masterpiece was *A Delineation of the Strata of England and Wales,*

Opposite Smith's Geological Map of England (which also covers Wales and part of Scotland).

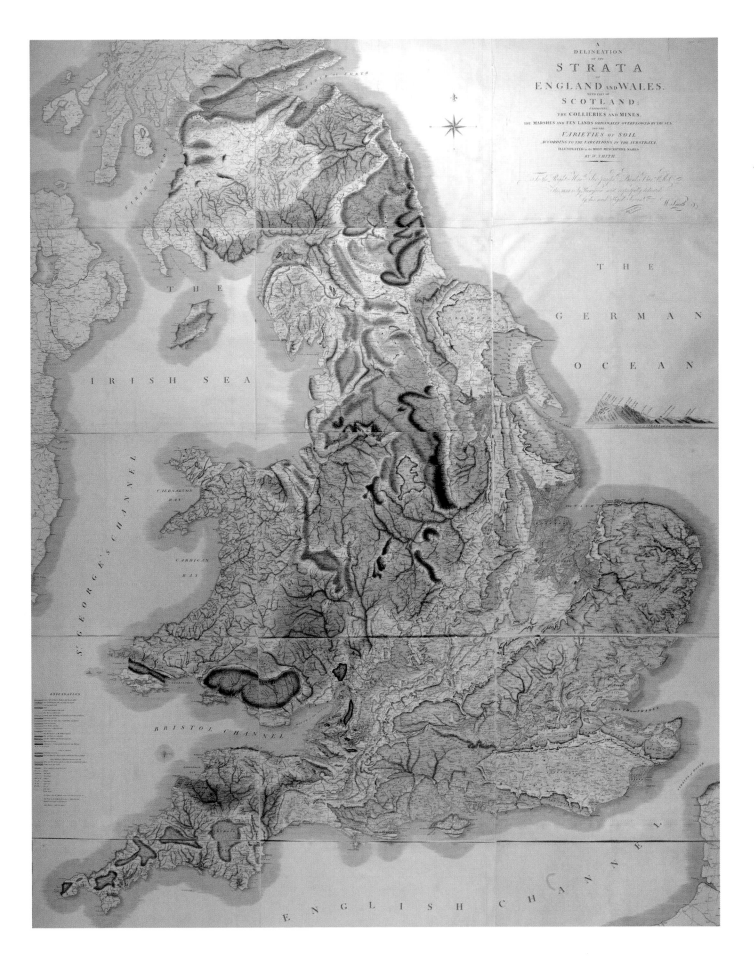

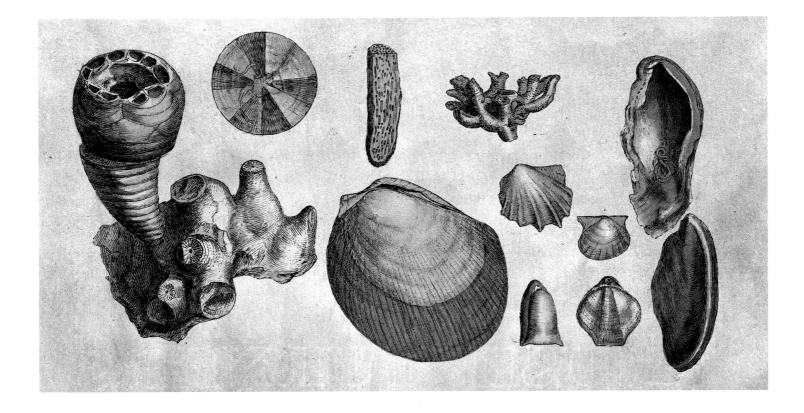

with Part of Scotland; Exhibiting the Collieries and Mines, the Marshes and Fen Lands Originally Overflowed by the Sea, and the Varieties of Soil According to the Variations in the Substrata, Illustrated by the Most Descriptive Names.

Drawing from his expertise as a surveyor, Smith intended the map to have widespread practical applications. Farmers, miners and more could use his map to work out where the best places might be to grow crops or mine coal. But Smith's map did more than that. The great map of his own country provided the background for England's geologic origin story.

Avalonia

Rocks of many different ages are preserved in England, including formations that gave western researchers their first looks at the life on the ancient past, but, as Smith's successors have found, much of England and Wales is made up of a lost microcontinent called Avalonia. The roughly 750 million-year-old rocks are testaments to the Neoproterozoic foundation of the area.

The origin of Avalonia goes back to ancient volcanoes. At a crease in the Earth's crust where one tectonic plate was being slowly shoved beneath another in the process called subduction, magma seeped to the surface to create islands in the prehistoric ocean. The movement of the plates caused the volcanic hotspot and the new rocks it formed to shift away from each other with time, creating an arc of volcanic islands. These islands coalesced into a microcontinent,

which continued to move with the Earth's plates as ancient Avalonia became part of the core of the supercontinent Pangaea.

But what is brought together is often torn apart again. As Pangaea broke up, the remnants of Avalonia were taken with it. A great ocean rift – the early Atlantic Ocean – shunted parts of Avalonia to either side, becoming parts of eastern North America, southern Ireland, and southwest England. These rocks form the foundation upon which many others rest, stone formed from the Earth itself that has been pushed, pulled, squashed and torn, an ancient remnant from a world just before the dawn of animal life.

Top Drawings by Smith of fossils sorted according to the strata in which they were found.

Above Carboniferous conglomerate rock studded with garnets, dating from Avalonia.

Opposite A drawing by Smith of a fossil mastodon tooth he found in Whitlingham, Norfolk.

Snowball Earth

Earth, the Blue Planet... When we think of our home, whether now or in the past, we generally imagine a vibrant and varied place teeming with life in azure ocean and on green land. But about 650 million years ago, around the time that the first animals on Earth evolved, the world may have turned into an immense ball of ice.

Experts call this the Snowball Earth hypothesis. It is proposed that, long after cyanobacteria oxygenated the Earth's atmosphere, around the time early animals like sponges had evolved, the temperature of the entire planet plummeted. Glaciers extended into the tropics, and most of the oceans were covered in ice. The whole world was as cold as modern-day Antarctica.

Unexpected glaciation

The key pieces of evidence for a Snowball Earth come from evidence of glaciers where there shouldn't have been any. As early as 1871, geologists were finding that immense ice sheets once covered and moulded ancient environments that were previously thought to have been warm. Much of the evidence for this glaciation came from tillite – little crumbs of rock that were broken off and transported by ice, often carried far from where they originated.

One of the places evidence was found was Reusch's Moraine, in Norway, named after its discoverer, Norwegian geologist Hans Reusch. The rocks deposited here were formed during the Neoproterozoic, over 541 million years ago, with tillite laid on top of sandstone. The sandstone must have been the foundation for a vast amount of ice. Striations along the sandstone's surface indicate that ice scarred the landscape, as glaciers have done in other times and places.

Above Geologist Hans Reusch (1852–1922), left, alongside his colleague Waldermar Brøgger with the tools of their trade.

Opposite The rocks in Flinders Ranges in South Australia show evidence of glaciation in the Neoproterozoic Era.

Ice in Australia

When Reusch documented this place in the late nineteenth century, the tillite just seemed like an oddity. The concept that the entire world might have frozen over didn't gain traction until the mid-twentieth century, when Australian geologist Douglas Mawson identified evidence of glaciers in the Neoproterozoic rocks of South Australia. Mawson's colleagues were not convinced. Not long after Mawson's proposal, however, British geologist W Brian Harland revived the idea that the Earth was once very icy. Harland examined when and where tillites from Greenland and Svalbard formed, discerning that the ancient rocks in these places had once formed in the tropics – and that they showed the mark of ice.

Albedo feedback

American geologist Joseph Kirschvink coined the term 'Snowball Earth' in 1992 to describe this time in our planet's history. According to the idea's adherents, the events that led to the global freeze went something like this. Ice isn't just chilly; it is also very good at reflecting light – and hence heat – from the sun. Researchers call this ice-albedo feedback. The more ice there is, the more of the sun's energy and heat is reflected back out into space. This creates a cooling effect that will cause temperatures to drop unless there is a mitigating force like volcanic activity spewing greenhouse gases into the air. The cooler the global temperatures get, the further the glaciers expand. Researchers estimate that, by the time ancient glaciers had extended to within 25 degrees latitude of the equator, the feedback loop would have been powerful enough to cause ice at the equator.

Ice-scratched stones from about 650 million years ago are the critical piece of evidence for this scenario. Even recognizing the movement of continents through time thanks to plate tectonics, there seem to have been warm places where there are signs of ice. On top of that, some geologists have proposed that banded-iron formations made a reappearance at this time. If that's so, then the banded stone indicates a drop in the dissolved oxygen contained in the ocean – a condition that might have been brought on by ice-covered seas in which cyanobacteria couldn't photosynthesize, causing an oxygen slowdown that was followed by an increase when conditions eventually thawed.

Disputed evidence

Not all experts are swayed by this scenario. Rocks of the appropriate age are very difficult to find and correlate with each other. Unless they can be accurately dated to the same time, the stones of 650 million years ago might document shifting conditions over thousands or millions of years. And, critics point out, many of

the striations and dropped stones attributed to glaciers may have been created by other means. The rock record can be tricky, and appearances do not always match up with a proposed hypothesis.

Strangely, the fossil record from this time does not indicate a mass extinction. During the appropriate time window – known to experts as the Cryogenian – there is no sign that the organisms we know about died back or were constrained by a deep freeze. Perhaps such life forms were not as affected by cold or found refuges, such as around hydrothermal vents. Or it may be an indication that the Snowball Earth was more of a slushball, or that some experts have misread the rock record. For the moment, the answers are still kept in the ancient stone.

Above Diamictite in Virginia, USA. Formed during Snowball Earth, this glacial conglomerate contains fragments of 1.3-billion-year-old granite.

Opposite Sea ice reflects sunlight, contributing to global cooling through ice-albedo feedback.

The Ediacara Hills

If you want to get a good look at early life on Earth, there's no better continent to search than Australia. Not only does it host ancient stromatolites, but the Ediacara Hills north of Adelaide also preserve a record of evolution's wild phase during the early years of animal life.

So far as scientists have been able to tell, the first animals evolved about 700 million years ago. These were not like the mammals, insects, fish or sea anemones around today, though. Animals are defined as multicellular organisms that do not make their own food. They are organisms that are made up of varied component cells that perform different functions. This is what makes a sponge an animal even if it doesn't move around a lot and lacks features like eyes or limbs. Animals can come in many different forms.

Strange forms

By about 575 million years ago, early animals had moved far beyond balls of multiple cells to comprise a menagerie of bizarre forms. We know this from the fossils found in the Ediacara Hills. Some are related to animals alive today; others are so strange that they seem to defy classification.

No one expected to find this Ediacaran garden, whose discovery was an accident. In 1946, Australian mining geologist Reginald Sprigg was poking around the rocks outside Adelaide when he

Right The worm-like *Spriggina* is one of the many animals from the Ediacaran biota that have resisted classification.

Opposite Life in the Ediacaran sea may have included plant-like animals that were fixed to the seabed.

found strange impressions in the ancient Precambrian sandstone. What these fossils were wasn't immediately clear. Many were flattened, pancake-like impressions or came in frond-like shapes, drawing comparisons with jellyfish and worms.

Mysterious fossils

One of the most common fossils found in the area, first described by Sprigg in 1947, is *Dickinsonia*. Imagine an earthworm squashed flat into a round blob ranging from a few millimetres to over a metre wide and you've got the picture. For years, paleontologists weren't quite sure how to classify it or even what its anatomy was really like. Different experts proposed that it might be a fungus, a jellyfish, a lichen or a member of a totally lost kingdom of organisms. More recently, geochemical analyses have found that *Dickinsonia* fossils contain traces of cholesterol. This is a biomolecule made only by animal cells, so its presence indicates that the strange creature is part of the same kingdom that we belong to.

Other organisms found in the same rocks have been harder to interpret. Consider *Spriggina*, an organism named in honour of

Sprigg himself. The fossil has a curved portion that seems to be some sort of head or anchor for a longer, frond-like part. Looked at one way, *Spriggina* is like a worm with a head shield. Looked at another way, it looks something like a sea lily or other type of animal that anchors to the seabed. Experts have settled on the idea that *Sprigginia* was an animal, but so far no one has been able to confidently discern whether it's an ancient relative of more familiar animals or belongs to a totally extinct group. Other fossils found in the same place, such as the disc-like *Tribrachidium*, have also resisted classification, awaiting more fossils and analyses.

Together, all the animals from this particular part of time are known as the Ediacaran biota. Not all the fossils from this group come from southern Australia. In the forests of England, for example, paleontologists have found a frond-like animal called *Charnia* in rocks of a similar age. But the Ediacara Hills are one of the richest hunting grounds for paleontologists focused on early life, and the evolution of this community marks a significant point in life's change from a purely microbial world to one with varied, disparate forms of life.

Explosion of species

The big change may have had something to do with the Snowball Earth hypothesis. If much of the Earth was covered in ice, perhaps to the point that photosynthesis in the seas slowed, then the retreat of the massive glaciers would have dramatically altered ocean chemistry. Early animals, in their primordial multicellular forms, had already evolved during the encroachment of the ice, and the shift back to warmer, oxygen-rich conditions might have opened new opportunities for animals to develop new niches. This ecological back-and-forth might explain why the Ediacaran biota pops up so suddenly, part of what some paleontologists have called the Avalon Explosion.

Unlike the stalwart stromatolites or other ancient organisms, none of the Ediacaran life forms has survived. By the beginning of the Cambrian, 541 million years ago, the likes of *Spriggina* and *Dickinsonia* were gone. Perhaps the disappearance of these organisms came through evolution, as those Ediacaran species evolved into the new forms that took off in the Cambrian. Or perhaps the Ediacaran animals we know about were part of an early evolutionary burst that was ultimately replaced by different branches of animal species. We don't yet know. Whatever the case, these ancient oddities marked the beginning of an evolutionary riot that continues through to the present day.

Above A fossil of *Dickinsonia*. The presence of cholesterol suggests that this was an animal.

Left Brachina Gorge in the Ediacara Hills is an area that is particularly rich in fossils.

Burgess Shale

High up in the mountains of British Columbia, Canada, lie the remains of an ancient seabed that is positively littered with fossils. The Burgess Shale is the type of fossil wonderland paleontologists dream about.

Some of the fossils in the Burgess Shale are familiar, such as the numerous trilobites that originally brought attention to this area. But many others are so otherworldly that experts struggled for years to classify them. Here in the stone rests an entire zoo of animals with compound eyes, grasping appendages, pointed body armour, crushing mouthparts and more – the prime representatives of the period that experts have come to call the Cambrian Explosion.

Hunting trilobites

No one really knows who was the first person to discover the Burgess Shale. Fossil collectors started bringing specimens back from this area in the 1880s. The word in scientific circles was that this spot was good for hunting trilobites – ancient and roughly bug-like arthropods that crawled along the seafloor and could roll up like woodlice when threatened. That scuttlebutt attracted American paleontologist Charles Doolittle Walcott to the area in 1909, but he found much more than trilobites. When Walcott and his family uncovered the fabled fossil beds, they started finding 508 million-year-old fossils that were anatomical head-scratchers. There were worms that weren't quite worms, jellyfish that seemed wrong for jellyfish, and total mysteries like *Anomalocaris* – the 'strange shrimp' that seemed to be a body without a head.

Walcott returned year after year to split the prehistoric shale and find the compressed creatures within. In all, he and his assistants collected more than 65,000 specimens. It was a collection so vast that Walcott himself never got around to studying and naming all the new animals he had found. When he did get a chance to describe them, he often associated them with what modern animal groups made sense. For instance, a little, soft-bodied squiggle of an animal called *Pikaia* was classified by Walcott as an ancient worm similar to an earthworm.

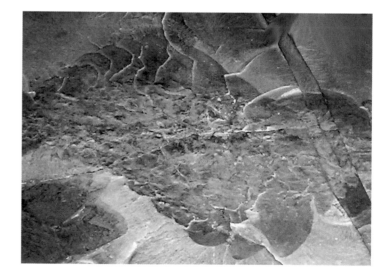

Above The first complete fossil of *Anomalocaris*, which confirmed that it was not a shrimp but a much larger invertebrate.

Opposite Fossil trilobites, which originally drew collectors to the area.

New classifications

Pikaia wasn't a worm. It was far more interesting than that. Later researchers found that this animal was one of the earliest chordates, or a precursor of animals with backbones. *Anomalocaris*, too, was not a shrimp. The first fossil to carry that name was actually part of a much larger animal whose various parts are often preserved separately. When a fossil of the whole animal was found, making sense of the strange pieces, *Anomalocaris* turned out to be a metre-long invertebrate with compound eyes, spiky appendages jutting from below its head and a shutter-like mouth.

Among the discoveries Walcott never got round to was an animal named *Hallucigenaia* because it looked like a hallucination. This was a spiky version of today's velvet worms, but its anatomy was so odd that paleontologists initially reconstructed it upside down. Another later find, *Opabinia*, was a distant cousin of arthropods, with five eyes, a trunk tipped with a claw and a body that looked like a lobster tail. When the first image of this animal was presented to a scientific conference, the paleontologists reportedly laughed at what they initially thought was a joke.

Below Charles Doolittle Walcott (1850–1927), searching the Burgess Shale with his son Sidney and daughter Helen in 1913.

Opposite A scene showing Burgess Shale animals in the ocean. In the centre is *Anomalocaris*. Top left is *Pikaia*. The spiky creatures at the bottom are *Hallucigenia*.

Ancient ancestors

The various creatures that Walcott and other researchers collected are the spectacular fossils of animals that lived around an ancient reef. The amazing preservation of these soft-bodied animals, which normally do not readily fossilize, came from an underwater cascade of fine mud that killed and preserved the creatures in this spot. And as strange as many of them may seem to modern eyes, paleontologists have found that many of these Cambrian weirdoes are distant relatives of animals alive today. *Anomalocaris* was an early arthropod – part of the group that includes insects and crustaceans – while *Pikaia* may be one of our ancestors. Even though there are strange animals like *Marella*, which looks like a tiny starship from a sci-fi movie, the Burgess Shale documents a time when animal life was rapidly becoming more complex and the groups of animals we see today were getting their start.

What caused this exceptional bloom of early life? Scientists have a few ideas, and more than one might be right. Oxygen levels in Cambrian waters rose, allowing animals to breathe more efficiently and grow larger. The amount of the mineral calcium also increased as rocks on the continents eroded and their contents were swept to sea, with the excess calcium providing the raw materials to construct hard body parts. The evolution of eyesight may have played a role, with early predators and prey starting an evolutionary arms race between hunters and the hunted. Small changes had major consequences for life, resulting in some of the first complex food webs we know about. And this was just the beginning.

Tongue stones

Fossils are facts of prehistory, all the bones, tracks, leaves, shells and other vestiges of ancient life. But the scientific understanding of fossils is only about two centuries old. For centuries before that, 'fossil' only meant something dug out of the ground. There was no connection to Deep Time, evolution or extinction, and a shark's tooth was mistaken for a 'tongue stone'. Opinions started to change when fossil teeth were compared to teeth from today.

To be clear, the confusion was primarily among the scholars of Europe, who would eventually lay the groundwork for science and its philosophical underpinnings. Many of these early naturalists had a difficult time reconciling what they read in scripture with the facts of nature. Other people did not have the same struggles. Various indigenous cultures in North America, for example, correctly recognized that fossil bones and footprints were the remains of organisms that had lived and died long ago. Fossils became parts of stories and cultural identity, all through the understanding that whatever had left the fossils behind was once alive.

People have been finding fossils and pondering their meaning for at least 100,000 years. That's the age of a stone tool with a fossil echinoderm inside, one of many that seem to have been intentionally made through time. But as ideas about the age of the Earth and the history of life changed in different cultures, the meaning of fossils became obscured by myth and superstition. In seventeenth-century Europe, for example, fossils were not thought to be connected to prehistoric life at all.

Myth and meaning

There was no consensus about what the ancient shells, teeth and other fossils really represented. First-century CE Roman naturalist Pliny the Elder proposed that objects that looked like fossil teeth actually fell from the sky during lunar eclipses. Later scholars recognized that shells and teeth looked like their counterparts in modern organisms, but claimed that they were actually fakes. Fossils, in this view, were attempts by the Earth to imitate life. It was easier to believe that a fossil bone was created by the soil than to reconcile fossils with prehistory. Another proposal centred on fossil shark teeth, asserting that these serrated triangles were the petrified tongues of serpents turned to stone. These *glossopetrae,* or tongue stones, were thought to carry magical properties that could counteract poisons or toxins.

Opposite Nicolas Steno's 1668 drawing of the head and teeth of a great white shark, *Carcharodon carcharias.*

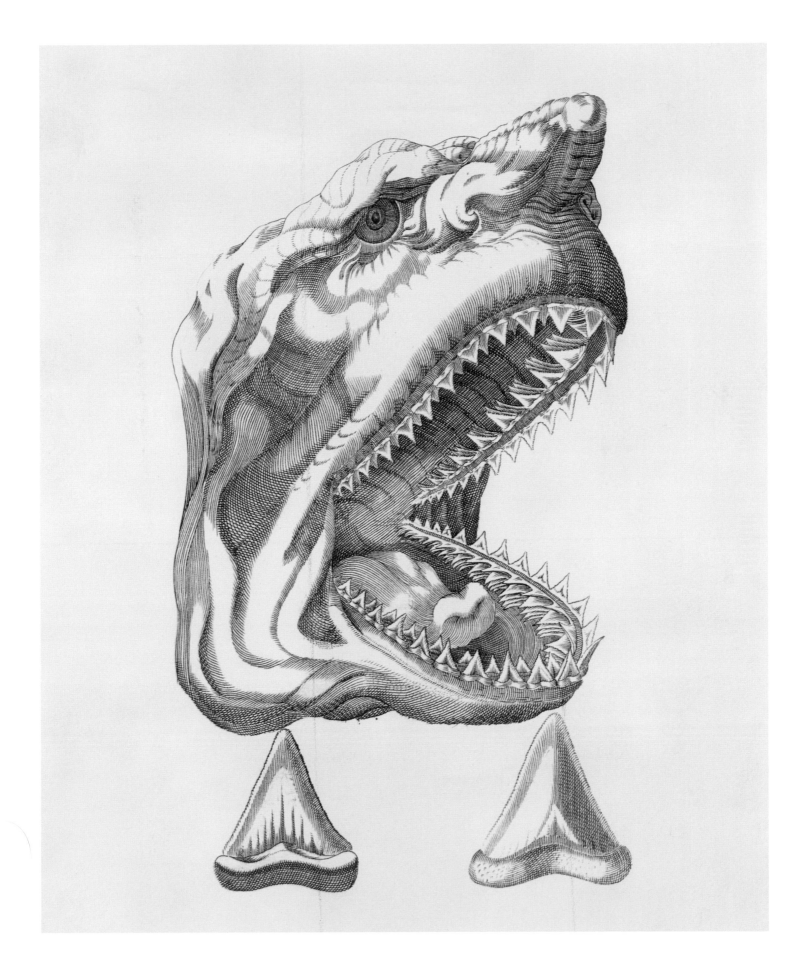

Above Fossilized shark teeth dating from the Upper Cretaceous.

Opposite A great white shark bares its fearsome fangs.

A great white shark caught off the coast of Livorno, Italy, in 1666 was the necessary catalyst to change the narrative. The shark's head, brimming with triangular teeth, was brought to Florence for examination by Danish anatomist Nicolas Steno. In picking over the desiccated head, Steno realized that those large teeth looked familiar. They were very similar to *glossopetrae*, to the point that a fossil shark tooth was illustrated next to a drawing of a fresh one in Steno's description of the shark head.

Rethinking geology

The connection between the shark head and the tooth-like objects perplexed Steno. He couldn't deny that the *glossopetrae* were, in fact, teeth from sharks that must have lived a long time ago. But how could he provide evidence for this? Steno began thinking about the details of geology, leading to questions about how old the teeth were and how fossils that would have been deposited in the seas could have wound up in rock on dry land.

The shark teeth must have been deposited on a seafloor and then covered by sediment, but how could fossils from sea animals wind up on land? To explain this, Steno outlined a principle that geologists now know as the law of superposition. In a typical set of strata that have not been tilted or otherwise altered, the older rocks are on the bottom and the younger ones are at the top. This follows from the way new sediment is laid down over existing sediment, creating layers over time. The shells and teeth found in European mountains were laid down in ancient seas, with land and water changing over time to eventually bring the fossil layers high above the surface of sea level.

Even though Steno's idea wasn't immediately a hit, the book he wrote on the subject inspired other naturalists, who picked up the thread. There were still more big ideas to discover – such as the realities of extinction and evolution – but, by the end of the seventeenth century, the reality of fossils as records of ancient life could no longer be denied.

Ordovician–Silurian Extinction

There are times in the history of life when the extinction rate spikes – crises and catastrophes wipe out huge numbers of species in a narrow window of time. These are called mass extinctions, and there have been five so far. The first occurred 440 million years ago, and is known as the Ordovician–Silurian Extinction.

Extinction is the counterpoint of evolution. Every species that arises anew eventually disappears. It is either extinguished or supplanted by descendant species. That's why paleontologists sometimes state that 99 per cent of all species that have ever existed have become extinct. The rate at which species tend to disappear over time is called background extinction. During the heyday of the dinosaurs, for example, individual dinosaur species tended to stick around for about a million years before they either disappeared or left descendants in the form of a daughter species. Mass extinction events involve disappearances that are faster by many orders of magnitude.

Ordovician fauna

The world paleontologists perceive when they look into Ordovician rocks seems as alien as another planet. There were no animals living on land. In fact, for most of the time period, there were no land plants either. Animals were entirely confined to the oceans, and invertebrates ruled. Following the Cambrian Explosion, entire communities of animals formed underwater ecosystems rife with trilobites, shelled animals called brachiopods, early fish, primitive

cephalopods and more, swimming over archaic corals and frond-like organisms called crinoids.

Even from space, Earth looked very different. Most of the world's terrain was grouped together in a massive supercontinent called Gondwana. This supercontinent straddled the Southern Hemisphere and gradually shifted southwards throughout the Ordovician. And that might have spelled bad news for life on Earth.

Cooling then warming

Something as big as a supercontinent stretching over the South Pole had major consequences for the planet. The Ordovician world was warm, but as Gondwana slid over the South Pole, the climate suddenly cooled. Massive glaciers started to form around Gondwana, as seawater turned into ice. Sea levels dropped as the

Opposite Top The shallow Ordovician seas were rich in life.

Opposite Bottom Fossils of *Dikelokephalina,* a trilobite that was widespread during the Ordovician.

ice built up, which was very bad news for the various organisms that lived in warm, shallow, sun-lit waters around the world. Then, almost as rapidly as the global climate had cooled, it warmed again. The rapid warming of the oceans led to waters that were depleted in oxygen while also being temporarily toxic to life.

This cold and hot flash was difficult for many forms of life to cope with. The Ordovician–Silurian disaster was a mass extinction that played out in two pulses about a million years apart, accounting for the loss of about 85 per cent percent of known marine species. This was the first time in Earth's history that biodiversity had suddenly plummeted.

Surprisingly, however, this mass extinction event did not seem to change much. All the major groups of animals that thrived

during the Ordovician – the corals, fish, trilobites, cephalopods and more – had members that survived into the next period, called the Silurian. Indeed, what is most strange about the Ordovician–Silurian Extinction is the way it did *not* fundamentally change the nature of the oceans, at least not right away.

During mass extinctions, severe cutbacks to biodiversity often allow new forms of life to thrive when the established orders are weakened or disappear. The mass extinction at the end of the Cretaceous, for example, eliminated all the non-avian dinosaurs and allowed mammals to move into new niches. But after the Ordovician-Silurian Extinction, ecosystems did not radically restructure themselves. Even though some groups of animals were totally lost – such as trinucleid trilobites, which looked like

miniature spaceships, and a heavily-ridged brachiopod genus called *Plaesiomys* – enough species survived from each major group to re-establish life in the seas.

Adaptable survivors

Most of the survivors were generalists. They could tolerate broader temperature ranges and eat more kinds of food than specialized species that were restricted to certain conditions or food sources. Some groups of animals never quite bounced back. Trilobites were incredibly abundant and diverse during the Cambrian and Ordovician, but their diversity was cut back by about 70 per cent during the extinction. Trilobites still existed, but they never returned to their previous numbers.

In time, life crept back. Instead of pockets of highly specialized and unique species, Silurian seas often contained broad-ranging species that evolved from the extinction-survivors. In time, these generated daughter species that proliferated throughout the planet. About five million years after the mass extinction, the number of species present in the seas had regained its losses. For a first brush with a mass extinction, life on Earth got incredibly lucky.

Above A well-preserved fossil eurypterid, an arthropod that survived the extinction to become abundant during the Silurian.

Opposite A eurypterid swimming in the Silurian ocean. This now-extinct group included the largest arthropods ever to have lived, with some species growing up to 2.5 metres long.

Coelacanths

Charles Darwin was a very prescient naturalist. He not only realized that the mechanism of natural selection could explain transcendent change among organisms, but he also understood that the same phenomena explained why some species didn't change much at all. One of the most remarkable examples of an unchanging species is a fish called the coelacanth.

In Darwin's famous book *On the Origin of Species*, he suggested that seemingly ancient species such as the platypus and lungfish might persist today because they lived in environments that had remained unchanged and therefore had not required dramatic adaptations. Animals like these, Darwin suggested, "may almost be called living fossils". Even so, biologists have often been perplexed by living animals that closely resemble those found in the fossil record. Stranger still, some creatures that they thought had gone extinct long ago appear to be alive and kicking. That's what shocked scientists about the coelacanth, a fish that was supposed to have been extinct for 66 million years.

Chance discovery

Two days before Christmas in 1938, South African trawler captain Hendrik Goosen returned to harbour with the day's catch. Goosen had a friendly relationship with local museum curator Marjorie Courtenay-Latimer, so he phoned the scientist to see if she might want to look over his catch. In particular, Goosen told Latimer that the trawl nets had pulled up a very strange fish unlike anything the fishers had seen before. The crew had set it aside for the scientist's inspection, and it's a good thing they did. When Latimer saw the fish, she knew it was special – she later described it as the most beautiful fish she had ever seen.

Latimer struggled to get other researchers to acknowledge the odd fish. She had it taxidermied to save it for science. Eventually fish expert JLB Smith confirmed Latimer's hunch. The slippery swimmer was a coelacanth, a very ancient form of fish that was thought to have gone extinct around the same time as *Tyrannosaurus*. The trawler may as well have pulled a dinosaur from the depths.

The heyday of coelacanths was deep in the past. The first of their family evolved at the beginning of the Devonian Period, about 390 million years ago. They were very different from the fish you might find in a pond. Coelacanths belonged to a group of fish called sacropterygians. The fish in this group have fleshy fins, like a mitten of muscle and skin over a core of articulated bones instead of the membrane stretched over fine rays like in a trout or salmon. In fact, we shared a common ancestor with coelacanths prior to 390 million years ago – our arms and legs are modified versions of the meaty appendages that typify these fish.

Diverse group

For tens of millions of years, coelacanths thrived in the oceans and fresh waterways of the world. During the time that the first dinosaurs were evolving, about 235 million years ago, coelacanths hit their peak diversity. Some were small enough to hold in your hand.

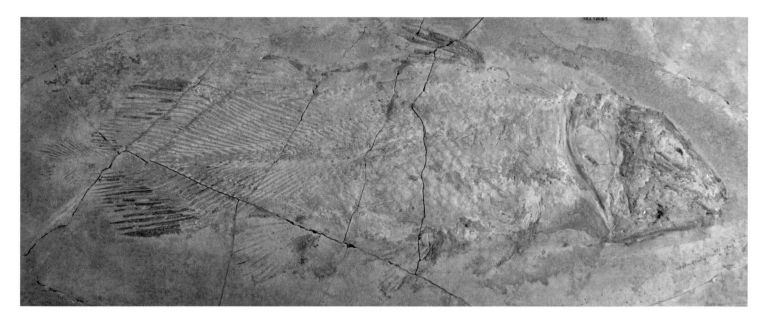

Above A fossilized *Undina penicillata*, a coelacanth that lived during the Jurassic Period.

Below The skeleton of a present-day species of coelacanth, *Latimeria chalumnae*.

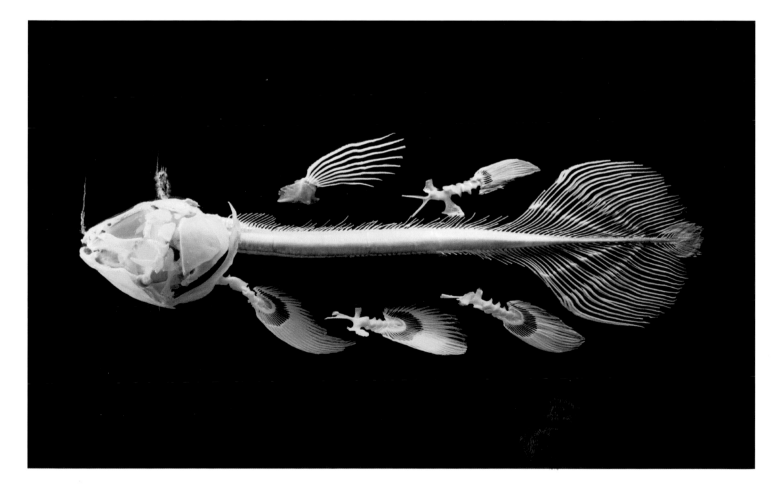

Others stretched 4 metres in length. By the end of the Cretaceous Period, however, coelacanths had all but disappeared.

No one expected that coelacanths had survived because their fossil record totally evaporated. Even to this day, nobody has found a definitive coelacanth fossil from the past 66 million years. Paleontologists have hypothesized that the apparent decline in coelacanths is merely an artefact of the fossil record. If coelacanths moved from shallow rivers and shorelines to deep water, the chances of finding their remains would have decreased. Deep-sea deposits are extremely hard to find in the fossil record, which could have obscured the fact that these fish survived.

In honour of Latimer's discovery, Smith named the first living coelacanth species *Latimeria chalumnae*. The fish lives in the deep waters off South Africa, Madagascar and Mozambique, growing as long as 2 metres in length. And while you might have seen ghostly specimens in museum display cases, the living fish is a resplendent blue dotted with white.

Second species

Perhaps more surprising than Latimer's initial discovery was the fact that more than one coelacanth species has made it to the present day. On 18 September 1997, biologists Arnaz and Mark Erdmann were travelling in Indonesia when they noticed an odd fish in the Manado Tua market. The couple eventually posted their photos on the internet, which caught the attention of a fish expert, who recognized that this was a unique species. It was named *Latimeria menadoensis* in 1999.

Despite their reputations as living fossils, however, today's coelacanths are not carbon copies of their Devonian forebears. They are adapted to life in the deep sea rather than the shallows, and now that scientists can sequence coelacanth genes, researchers have found that different coelacanth populations have varied genetic markers – the fish are still evolving. Sometimes slow and steady wins life's race.

Left An artist's rendition of a coelacanth swimming along the ocean floor.

Late Devonian Extinction

Paleontologists started detecting mass extinctions before they truly understood these devastating events. One of the most puzzling of all appears to have taken place in a number of distinct pulses about 376–360 million years ago. Known as the Late Devonian Extinction, its causes and meaning have been hotly debated.

As experts catalogued the world's strata and the creatures within them through the nineteenth and early twentieth centuries, there seemed to be definitive breaks during which life on Earth rapidly changed. Some suggested that vast cataclysms had shaken up life's order, favouring some organisms over others. Others said that these events were just artefacts of an incomplete fossil record and that such shifts were actually gradual. Then there were those who saw all these changes as signs of progress, of life reaching higher and higher levels of complexity and advancement through an Age of Fishes, Age of Reptiles and Age of Mammals, through to our current age where humans dominate.

In time, paleontologists came to realize that these distinctive breaks were not phantoms of the fossil record. Sometimes extinction strikes down entire groups of organisms in a geological instant. However, these events don't always happen at boundaries between great geological periods, like the Orodovician-Silurian catastrophe (see page 84). Sometimes mass extinctions strike seemingly in the middle of a geological period, as was the case with the Late Devonian Extinction.

Age of Fishes

While the term isn't the most scientific – there were many other forms of life around at the time – the Devonian could fairly be called the Age of Fishes. Fishy ancestors, such as the little swimmer *Pikaia* in the Burgess Shale, had been around since the time of the Cambrian Explosion, but early vertebrates were kept somewhat to the sidelines for tens of millions of years. The cephalopods, trilobites and the strange eurypterids (also called sea scorpions, see page 87) were the major players in the seas. During the Devonian, however, starting 419 million years ago, fish evolution truly took off. The seas filled to the brim with ancient relatives of sharks, bony fish, coelacanths, lampreys and many more, including the ancestors of the vertebrates, which would eventually evolve limbs and drag themselves out of the swamps to make a home on land.

Ancient forests

The supercontinent Gondwana was still the world's major landmass during this time, stretching over a great deal of the Southern Hemisphere. There were other landmasses, such as an ancient continent called Laurussia that sat on the equator, but Gondwana comprised most of the world's land. There were many plants there. The little algae and plants that could tolerate life out of water during

Opposite Top A typical Devonian reef system, featuring the placoderm *Dunkleosteus* (top centre), *Pterygotus* (top left), *Bothriolepis* (bottom right), ammonites (top right) and *Stethacanthus* (centre left).

Opposite Bottom *Pteraspis* (left) was a genus of jawless fish that went extinct in the Late Devonian.

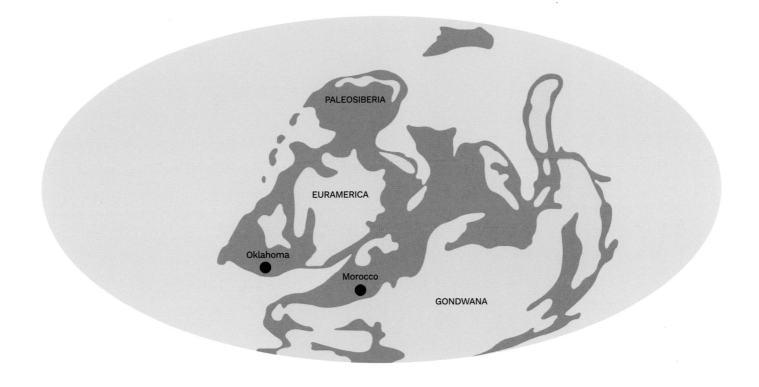

PALEOSIBERIA

EURAMERICA

Oklahoma

Morocco

GONDWANA

the Silurian crept further onto land and started to form ancient, towering forests. Insects found all this green food too good to resist, giving them the impetus to follow onto land and provide live bait for early four-legged fishes at the water's edge.

But something happened at the beginning of the last subdivision of the Devonian, a slice of time between the earlier Frasnian and later Famennian chapters of the period. The story of this extinction spans millions of years, as there seemed to be an uptick in the background extinction rate for about 20 million years prior to what paleontologists call the Kellwasser event, 372 million years ago, when oxygen levels in the ocean seemed to plummet. Then, 359 million years ago, there was another extinction spike called the Hangenberg event, which affected both life in water and life on land. This biodiversity crisis wasn't as stark as would be caused by an asteroid hitting the Earth. Rather, it was a protracted process in which species perished faster than they were being replaced. The world lost some forms of life forever during this period, including the placoderm, an early jawed fish covered in plates of bone.

Falling oxygen

The long stretch of time over which these extinctions played out has made them difficult to study. During the earlier phases of the extinction, ocean oxygen plummeted. Geologists find what are called anoxic shales from this time. These are rocks that show few signs of life. They were made of ancient seafloor mud in which there was little oxygen. Jawless fish, which had been prolific prior to this period, suffered as they had specialized in sucking up debris from the bottom.

Life itself might have had an active part to play in this process. The later part of the extinction affected life on land, and this may have been the result of early trees. Plants big enough to bust stone with their roots may have created more loose soil that then washed into rivers and out to sea, flooding the water with nutrients. This would have created algal blooms, which in turn caused the oxygen crashes. Sometimes the evolutionary success of some is a drawback for many more.

Above The distribution of land during the Devonian, when it was concentrated in the Southern Hemisphere.

Opposite Fossil-bearing Devonian Old Red Sandstone rocks extend across the North Atlantic region, from the northeastern seaboard of North America in the west to Norway in the east.

Coal

Children are threatened with coal in their stocking if they've been naughty prior to Christmas Eve. But an aspiring young paleontologist might be thrilled to get such a gift. That's because much of the world's coal is a time capsule from an extraordinary era when insects grew to tremendous sizes and thick forests covered the land.

Y ou can guess the significance of the Carboniferous Period from its name. Stretching from 359 million years ago until 299 million years ago, this was a time when incomprehensible amounts of carbon were buried and transformed into coal. And this was not an abiotic happenstance of the stone. Ancient life had everything to do with the massive coal seams found in Carboniferous rocks around the world.

Giant flora and fauna

The Carboniferous was a time of dramatic change for life on Earth. The seas started to look a little bit more like they do today. There were still ancient oddities like trilobites and coil-shelled ammonoids, but there was also a profusion of fish. On land, the great forests that spread during the Devonian continued to grow and change – bizarre prehistoric jungles in which archaic plants such as tree ferns could grow as tall as redwoods and had scale-like bark on the outside. Some of these trees were growing so tall that plants had to evolve a new biological substance, called lignin, in order to build support tissues to keep from toppling over.

Life on land was getting a little more crowded, too. Invertebrates – including primordial cousins of dragonflies, millipedes, spiders and more – had been making their homes on land for millions of years by the beginning of the Carboniferous. Thanks to some early landlubbers that went after arthropod snacks near the end of the

Devonian, vertebrates were also becoming much more prolific. Many were amphibians, forms similar to great salamanders that could venture onto land but had to keep their skin and their eggs wet. But some of these creatures adapted to the drier conditions on land and became capable of laying eggs enclosed in waterproof shells.

Above *Pederpes*, a metre-long Carboniferous tetrapod.
Opposite Typical Carboniferous flora.

Above This fossilized Carboniferous insect had a wingspan of 68 centimetres.

Opposite An exposed coal seam from an old mine in Canada.

The shells prevented the developing embryos from drying out. These animals are called amniotes, and soon after the first amniotes started to make their home on land, the earliest forerunners of reptiles and mammals split from each other and began their independent evolution.

Logs piling up

This riot of life was largely made possible by the pioneering plants. Forests took root in what had previously been hard soil. These stands of primitive trees grew thick and tall. The lignin that allowed them to do so also had another effect. The tough material made it much harder for the trees to break down after they had died and been toppled. Bacteria capable of decomposing lignin had not yet evolved, or were not yet very efficient at it, so great logjams of these ancient trees piled up in the Carboniferous swamps.

The surfeit of vegetation, both living and dead, had two major consequences. The first has to do with what organisms with chloroplasts have been doing since the time of cyanobacteria – the plants produced copious amounts of oxygen. This many plants,

carrying out this much photosynthesis, flooded the atmosphere with oxygen. Instead of an extinction, though, this time organisms were capable of utilizing the boon. Insects, which respire by way of tiny holes called spiracles on their bodies, could breathe more efficiently and evolve to unprecedented sizes. The millipede-like *Arthropleura*, for example, reached more than two metres long and looked like a big, scaly area rug. Early amphibians benefitted, too, their ability to respire through their squishy skin allowing them to become numerous and very large – imagine a sharp-toothed newt about the size of a crocodile.

And here's where the coal comes in. Plants take in carbon dioxide from the atmosphere, using it to build their bodies. But the Carboniferous organisms were inefficient at breaking down these plants, so the downed trees and other green stuff deposited in the swamps didn't so much decay as wait for burial. Over time, with heat and pressure, it turned into coal. We release the carbon back into the atmosphere whenever we burn this fossil fuel. Human-made climate change is caused in part by burning the remains of prehistoric swamps.

The Great Dying

In the heart of South Africa, covering an expanse of 400,000 square kilometres, lies the Karoo. In this harsh, scrubby desert, valleys and escarpments present rocks from ages long past. This is the place many paleontologists come to when they want to understand one of the worst events in the history of life on Earth – the Permian–Triassic Extinction, or, as it is more notoriously known, the Great Dying.

In the Permian world, spanning 299 to 252 million years ago, protomammals were the dominant forms of life on land. These were our ancestors and relatives, fuzzy creatures that were once called 'mammal-like reptiles' for their mishmash of features. Some were hulking, pig-like herbivores with beak-like mouths and prominent tusks. Others were fleet-footed carnivores with sabre teeth. Complex ecosystems had evolved, with these animals holding a prominent place in the terrestrial realm.

Rise of the reptiles

The Triassic world was very different. During this time, between 252 and 200 million years ago, very few protomammals remained. The ecological torch had been passed over to reptiles, including early dinosaurs. The Permian closed the span of the Paleozoic and the Triassic kicked off the Mesozoic, split from each other by the worst mass extinction the world has ever seen.

All mass extinctions have been dramatic, but by body count, this third disaster was the worst. About 70 per cent of known species on land died, along with about 81 per cent of all known species in the seas. Diverse reef communities in the ocean and complex forest habitats on land were reduced to what paleontologists know as 'survivor taxa' – depauperate landscapes in which the relatively few surviving species spread everywhere. For example, the tusked protomammal *Lystrosaurus* was one of the few protomammals that survived the cataclysm. It spread through diminished habitats where little other than seed ferns grew during the immediate aftermath.

Above *Dimetrodon,* an Early Permian protomammal.

Opposite The extinctions were brought about by volcanic activity.

Volcanic eruptions

What could have so fundamentally damaged Earth's ecosystems while still providing a fresh start for the survivors? The answer comes from within the Earth itself. Intense and unprecedented volcanic eruptions rapidly changed everything from the atmosphere to the oceans.

Before the main event, there had been a lesser disaster. Around 260 million years ago, the Emeishan Traps in ancient China poured out large quantities of molten rock and greenhouse gases such as methane. These were not eruptions that sent lava cascading into the air, but more like suppurating wounds in the Earth that allowed the lava to pour forth. This particular eruption pulse seems tied to the disappearance of relatively archaic protomammals – similar to the sail-backed *Dimetrodon* – but did not cause widespread habitat destruction.

The situation was made far worse by a second, even grander eruption of the Siberian Traps in what's now Russia. These volcanoes, too, seeped and gushed, covering about 2 million square kilometres in volcanic rock. That's a huge area of land, but it wasn't the soupy volcanic output that nearly killed life on Earth. It was the aerosols and debris that these volcanoes put into the air, given an assist as the lava burned buried coal beds, releasing extra carbon dioxide.

Cascading catastrophe

The atmospheric output had several consequences. First, the dust was so thick that it blocked sunlight and temporarily halted photosynthesis. This was bad news both on land and in the oceans,

where photosynthetic plankton formed the basis of the ecosystem. Second, the greenhouse gases eventually dissolved in the ocean, turning the water slightly acidic. Shell-building and reef-building organisms, also important components of Permian ecosystems, found it much harder to build their armour and homes.

Even more seriously, the gaseous output of the eruptions was so great that the relative amount of oxygen in the air dropped. Organisms soon had their breaths stolen from them. Some life forms fared better than others. The protomammals breathed in a way similar to us, with deep breaths in and out, and they struggled for air. By contrast, many reptiles breathed more efficiently, with a unidirectional system in which incoming air helps push out air that has already been depleted of oxygen. (This is why birds today do so much better at high altitudes than we do.) The reptiles were pre-adapted to survive and thrive in the aftermath.

The effects were not instantaneous. The eruptions created a shifting set of circumstances that organisms struggled to adapt to. Even when the dust and debris from the eruptions settled, the greenhouse gases still drove a pulse of global warming that added pressure to the species that survived. Many of the rocks exposed in the Karoo span the worlds of the Permian and the Triassic. They record this changeover, when one world gave way to another.

Above An eroded rock formation in the Karoo desert.

Opposite A fossil of *Seymouria baylorensis*, a Permian amphibian.

Reptiles invade the water

When paleontologists talk about the Age of Reptiles, the focus often falls on terrestrial titans like the dinosaurs. Creatures like *Stegosaurus* and *Tyrannosaurus* would seem to be the epitome of saurian success. But during the same time, reptiles also thrived in the seas. They came in all shapes and sizes, from the ancestors of today's sea turtles to vacuum-faced algae-eaters unlike any species alive today.

Strange as it may be to consider, there were no ancient equivalents to whales, otters or seals prior to 250 million years ago. All the life in the seas, so far as we know, comprised creatures whose ancestors evolved in those environments and stayed there. Even as protomammals spread across the planet during the Permian, seemingly none of them ever became adapted to semi-aquatic or marine life. They remained purely terrestrial. It was only with the rise of reptiles that tetrapods reversed evolutionary course. Their ancestors had crawled out of the water, and now they were making a return.

Return to the ocean

The proliferation of marine reptiles in the seas had begun by 249 million years ago, which paleontologists consider to be a quick turnaround given that the Great Dying had shaken up the planet 'just' three million years prior. Fossils uncovered in southern China indicate that this was the time when the first marine reptiles became suited to life in the oceans.

What followed was an explosion of seagoing reptiles. In fact, the Triassic was the most diverse period in all of history for saurians that lived in the oceans. Some are connected to creatures around us today. The 220 million-year-old *Odontochelys* was an ancient turtle with a partial shell made of expanded ribs, ancient enough to still

have teeth in its mouth. Then again, species like the 245 million-year-old *Atopodentatus* were so strange as to seem alien to us. This marine reptile had a long tail, stubby limbs and a head shaped like an upside-down 'T' filled with tiny teeth, thought to be useful in scraping algae from nearshore rocks.

And there were far more. Some lineages, like the rotund placodonts, were shell-eaters that evolved during the Triassic but did not survive until the end of the period. The strange, long-necked *Tanystropheus* plied a similar path, swishing through the shallows after small morsels but lacking evolutionary staying power. Stranger still were the hupehsuchians, creatures that only existed within a million-year window. These reptiles look like they were put together by committee, with elongated jaws, paddle-like fins, tall spines on their backs and interlocking bones that served as a kind of body armour. Nothing like the hupehsuchians has evolved before or since, and paleontologists have taken their protective osteology as a sign that large marine predators also evolved early.

Opposite A skeleton of *Placodus gigas,* a genus of marine reptiles that swam in the shallow waters of the mid-Triassic oceans. *Placodus* fossils have been discovered in Central Europe and China.

The Triassic wasn't only a time of one-offs. The ancestors of the long-necked, four-paddled plesiosaurs, which would become a central part of the seas during the Jurassic and Cretaceous, got their start in the Triassic from an early group of similar reptiles called pistosaurs. Even more prolific were the ichthyosaurs, or 'fish lizards'. The earliest ichthyosaurs were almost eel-like in form. They probably swam like sea snakes do today. Over time, however, ichthyosaurs evolved the ability to paddle with their half-moon-shaped tails and steer with their fins, as sharks do, and some became apex predators. By the end of the Triassic, the mighty *Shonisaurus* had appeared. This ichthyosaur was the size of a humpback whale and had a mouth full of teeth each larger than your thumbnail. It evolved to eat other marine reptiles or whatever else it could catch.

Right place, right time

The question remains as to why reptiles so enthusiastically jumped into the drink. Even after the Age of Reptiles closed, the descendants of the protomammals never attained the same spread of different forms. There was something special about reptiles that allowed them to take to the water in ways that other vertebrates could not.

The earliest amniotes – the ancestors of both protomammals and reptiles – laid eggs to make a living on dry land. But eggs are not always deposited in the ground. As we know from many modern reptiles, some species keep the eggs inside their bodies until the little ones hatch and then emerge, a reptilian version of live birth.

Protomammals were likely capable of this kind of live birth, but for most of their history the seas were already full of creatures big enough to make a meal of any fuzzball that tried to swim. After the end-Permian mass extinction had cleared the decks, reptiles were able to venture into the water and plough new niches as the oceans recovered. The reptiles were in the right place at the right time, leading to an impressive splash of species.

Above The whale-sized ichtyosaur *Shonisaurus* hunting squid-like belemnites.

Dawn of the dinosaurs

The first dinosaur was not an imposing reptile. The earliest of the 'terrible lizards' was about the size of a German shepherd. The lanky creature had a small head, leaf-shaped teeth, a long neck, slender legs and a tapering tail, its body covered in a combination of scales and fluff. This creature was not the ruler of anything yet. In fact, for millions of years, dinosaurs were on the sidelines in a world dominated by their crocodile cousins.

During the Triassic, many different forms of reptiles developed new niches on land, in the seas and in the air, but the most successful group of all were the archosaurs. Their name translates to 'ruling reptiles', and for good reason. While the earliest members of their family evolved during the Permian, these reptiles were relatively marginal parts of ancient ecosystems until the end-Permian mass extinction. Their efficient respiratory systems allowed them to survive, and those survivors reached sexual maturity relatively early in their lives. That meant that the archosaurs were able to out-compete other forms of reptiles by laying large numbers of eggs with greater frequency, preventing the slow-reproducing protomammals from returning to their previous dominance.

Archosaur family tree

To understand who was found in the greater archosaur family, it is helpful to start from the present and work backwards. Today there are two living subgroups of archosaurs – birds and crocodilians. If we follow those two lineages back to their last common ancestor, every bygone lineage that falls between the bird and crocodile lineages is an archosaur. This not only includes dinosaurs, as birds are just one form of dinosaur, but also extinct groups such as the flying, leathery-winged pterosaurs that also evolved during the Triassic.

The crocodile side of the family tree did very well during the Triassic. These are known as pseudosuchians, but they were not like the aquatic ambush predators we know today. In the Triassic, crocodile relatives primarily lived on land and carried their legs beneath their bodies. Some were very small and lanky, almost like what you'd get if you crossed a crocodile with a greyhound. Others were terrible predators with massive, deep heads filled with serrated teeth. Still more were almost like early dinosaurs, having independently evolved the ability to run on two legs.

First dinosaurs

Early dinosaurs needed time to diversify into the array of forms that we've come to know and love. The earliest possible dinosaur we know of is called *Nyasasaurus*, a 243-million-year-old reptile recovered from the Triassic rock of Tanzania. The few bones we know from these animals bear a strong resemblance to those of another Triassic group called silesaurs. We know from fossil dung

Opposite A fossilized footprint of a crocodile relative found in Triassic sandstone in a quarry in northern England.

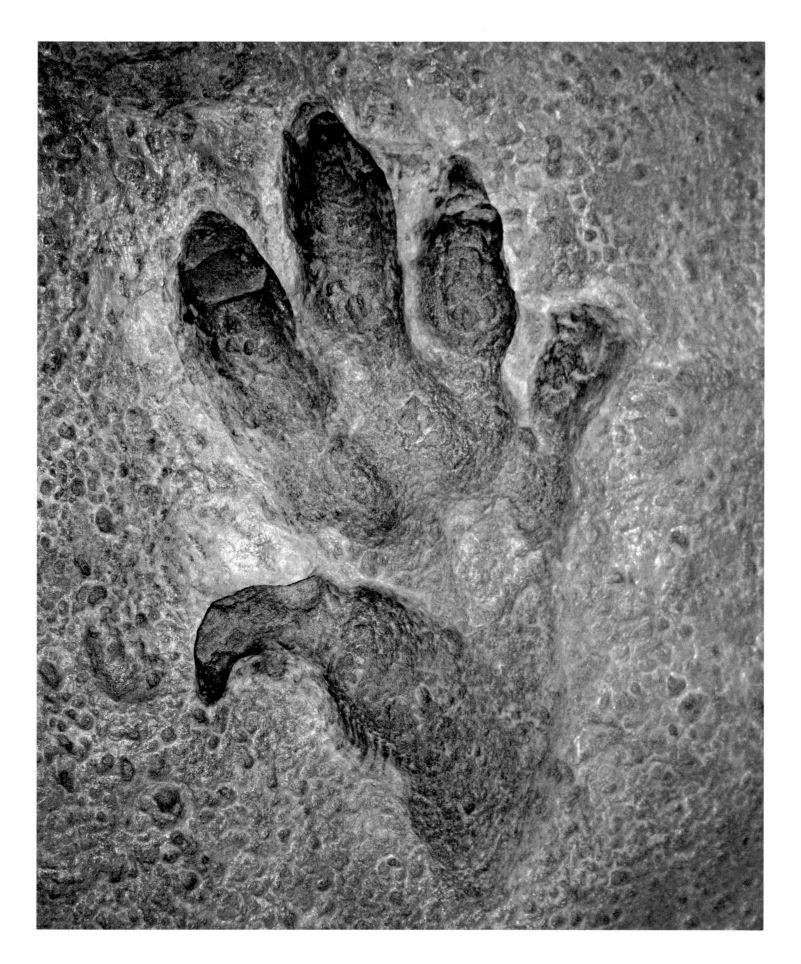

that they ate beetles and plants – ancient omnivores that lived alongside the dominant pseudosuchians. The first dinosaurs were a slight evolutionary modification of these ancestral reptiles.

Through the millions of years that followed, dinosaurs started to take up new niches. Some of them became increasingly predatory, pouncing on small lizards and protomammals. They evolved sharper, recurved teeth and stood up on two legs. These were the earliest theropods, the 'beast-footed' dinosaurs. Other dinosaurs ate more plants and started to grow in size, perhaps as protection against the predatory animals of the time. These were the sauropodomorphs – strange reptiles with small heads, extremely long necks, clawed

hands, and bodies often held up on two legs. They later came down to four legs as the largest species required more support for their bulk. Paleontologists hypothesize that another major dinosaur group called the ornithischians evolved during this time, too, although their fossils are incredibly rare.

Small numbers

Even as the great dinosaur families – the distant ancestors of *Tyrannosaurus*, *Apatosaurus* and *Stegosaurus* – became established, these animals did not immediately dominate their ecosystems. They are often rare and sparsely distributed. In the fossiliferous deposits

of the American southwest, for example, paleontologists only find small, carnivorous dinosaurs. Sauropodomorphs and ornithischians were around, but they are found in other ecosystems and often in small numbers. Even though it's technically true that the Triassic was the dawn of the dinosaurs, they were only part of a much broader story, and it took tens of millions of years for early dinosaurs to start evolving into the characteristic forms that fill museum halls today.

If the state of the Triassic world had gone on unabated, dinosaurs may have remained just one expression of a broader reptilian symphony. The pseudosuchians may very well have held sway. They seemed perfectly capable of filling the same ecological roles many early dinosaurs did. But something happened to give dinosaurs the upper hand – or, upper claw. Another mass extinction, the world's fourth, is what turned the Age of Reptiles into the Age of Dinosaurs.

Above A family of *Massospondylus* at a watering hole. This large sauropodomorph dinosaur lived in the Early Jurassic.

Opposite *Saurosuchus*, a large carnivorous pseudosuchian from the Late Triassic.

Triassic–Jurassic Extinction

Seen one way, mass extinctions are disasters. They are swift and deadly events in which many of evolution's offshoots are permanently cut down. But through every mass extinction the Earth has suffered, some species survived and thrived in the aftermath. In fact, one of our favourite times in the history of life on Earth wouldn't have come to pass without a mass extinction. The Age of Dinosaurs didn't just end with a mass extinction. It also began with one.

Reptilian life was thriving by the end of the Triassic, about 201 million years ago. Great marine reptiles sloshed in the seas. Crocodile cousins and early dinosaurs ran about on land. Toothy pterosaurs flitted and flapped through the air. Almost every niche occupied by mammals today was filled by reptiles during the Late Triassic. Then the world's biota was once again devastated by volcanic eruptions that chilled the world before turning up the heat.

Lava flood

The Late Triassic was the heyday of Pangaea, the great central continent made up of all the Earth's major landmasses. In the middle of Pangaea, between the prehistoric Americas and prehistoric Africa, sat what geologists call the Central Atlantic Magmatic Province, or CAMP. Much like the volcanically active areas that brought about the end of the Permian Period, these hotspots oozed flood basalts over vast areas of land. Geologists estimate that the eruptions at the end of the Triassic ended up covering about 11 million square kilometres. From Morocco to Brazil to the northeastern United States, remnants of these expansive basalts can still be found today. Put the pieces together and you get a map of a geological wound that bled through the centre of Pangaea.

Naturally, the habitats covered by the lava were devastated by the outpouring. But the most destructive effects came from the incredible amounts of ash, dust, carbon dioxide and sulphur dioxide that were released along with the molten rock. The effects were very similar to what had transpired at the end of the Permian. Some of the carbon dioxide and sulphur dioxide was taken up by the seas, and the sudden influx of these compounds caused the oceans to acidify. Shell-building organisms, from ammonoids to plankton, again struggled to build their shells.

The climate changed dramatically, too. Copious amounts of sulphur dioxide in the atmosphere likely triggered rapid global cooling. The warm Triassic world that was so kind to reptilian evolution came to a rapid close. Even though many Triassic reptiles were probably not cold-blooded, they still might have had body temperatures that fluctuated with the air temperature or lacked the insulation to remain warm through harsher cold snaps.

Opposite Molten rock from the volcanic eruptions formed basalt columns similar to these at Svartifloss in Iceland.

As before, the large amounts of carbon dioxide thrown into the air by the eruptions set the stage for global warming following the cold pulse. Those animals that were capable of dealing with the cold soon found themselves facing a world that was even hotter than it had been previously.

All of these changes, from sunlit seawaters to the interior jungles of Pangaea, resulted in the loss of many forms of life. Up to 34 per cent of known genera in the oceans died off, including some entire groups of organisms, such as the eel-like conodonts – a favourite of paleontologists in biostratigraphy given how prolific they had once been. On land, many of the crocodile relatives – such as armour-covered herbivores called aetosaurs and carnivorous, bipedal hunters called rauisuchians – died out entirely. The last, straggling protomammals died out, too, leaving behind the earliest mammals to carry on their evolutionary legacy.

Mass extinctions are severe events in which many groups of organisms lose part of their diversity, suffering a cutback even as some members of their lineage survive. But dinosaurs seemingly weren't fussed by the Triassic–Jurassic Extinction. Theropods, sauropodomorphs, and ornithischians all survived and entered the first days of the Jurassic without skipping a beat.

Protected by fluff

What allowed dinosaurs to be so resilient? The answer might lie in coats of fluff and fuzz. Paleontologists have been gathering evidence that early dinosaurs – as well as their flying pterosaur relatives – had bodies at least partly covered in simple, hair-like fuzz, or protofeathers. Much like down helps baby birds insulate their bodies against large swings in temperature, the insulating protofeathers of dinosaurs and the fur of early mammals may have allowed them to withstand the worst of the chill and avoid overheating when temperatures rose again. It was the scaly, crocodile-line lineages that suffered most, with a division between who survived and who went extinct played out on the basis of body coverings. Sometimes being fluffy has a real survival advantage, and dinosaurs used this lucky break to begin their rise to global prominence.

Above The scaly archosaur *Postosuchus*, one of the largest carnivorous reptiles during the Late Triassic, did not survive into the Jurassic.

Opposite An illustration showing a *Redondasaurus* skeleton lying in a dried-up riverbed. This large phytosaur was killed off during the Triassic–Jurassic Extinction.

The Snider-Pellegrini Map

There is a discovery that happens over and over again when children look at maps of the world. The shapes of Africa and South America seem to fit together. The two are immense, continental puzzle pieces that clearly match up with each other. But it took some time for geologists to agree on the significance of this match-up.

Of course, we now understand that Africa and South America really were connected in the deep past. Continental drift and plate tectonics are facts of Earth's history. Everything from the rocks of the continents to the fossils of creatures that once lived there reinforces the idea. But, strange as it may seem, it took scientists over a century to accept what we can so plainly see on a globe.

In 1858, French geographer Antonio Snider-Pellegrini published a book titled *La Création et ses mystères dévoilés* (*Creation and Its Mysteries Unveiled*). Among its contents were a pair of maps centred on the Southern Hemisphere. In one, rough interpretations of South America and Africa sit in their present positions on the globe. In the other, the two are connected to one another, with North America and Europe nestled against each other above. Snider-Pellegrini proposed that these continents were once connected, and he didn't base his conjecture on geographical shapes alone.

both Europe and North America were identical. This indicated that these continents were once connected. And if that were so, then South America and Africa would have been joined, too.

Snider-Pellegrini was not the first person to propose this idea. Three centuries earlier, map-maker Abraham Ortelius suggested that the Americas were once joined with Europe and Africa, only to be ripped away by floods and earthquakes. However, nobody took much note of Ortelius's idea. There was no reason to think that masses of rock as big as continents could move, and such conjectures obviously ran up against theological interpretations suggesting that the world was created fully formed.

The same was true for Snider-Pellegrini's suggestion. Even if other experts conceded that the continents had matching coastlines, there was no mechanism to explain such dramatic global change. As far as the fossil evidence went, many experts preferred to think that these organisms independently formed in multiple centres of creation or that

Supporting fossil evidence

In Snider-Pellegrini's estimation, South America and Africa were connected to each other during the Carboniferous. The proof, he wrote, came from plant fossils. Carboniferous plant remains from

Above German explorer and geophysicist Alfred Wegener (1880–1930).

Opposite Top Antonio Snider-Pellegrini's map showing 'Before the separation' and 'After the separation'.

Opposite Bottom Wegener's map of Pangaea.

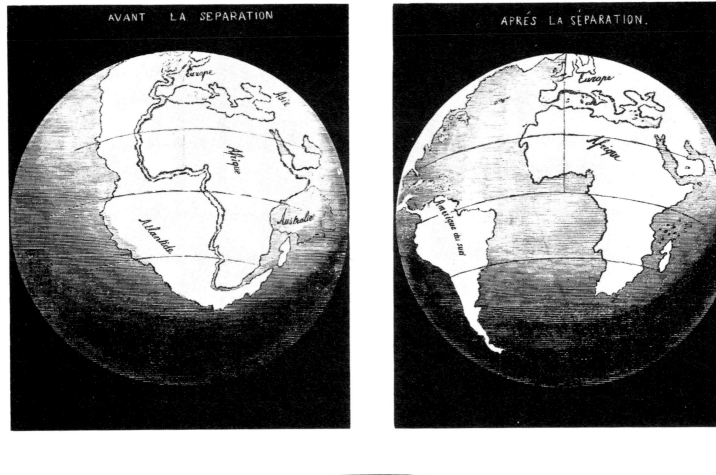

AVANT LA SEPARATION

APRÉS LA SÉPARATION.

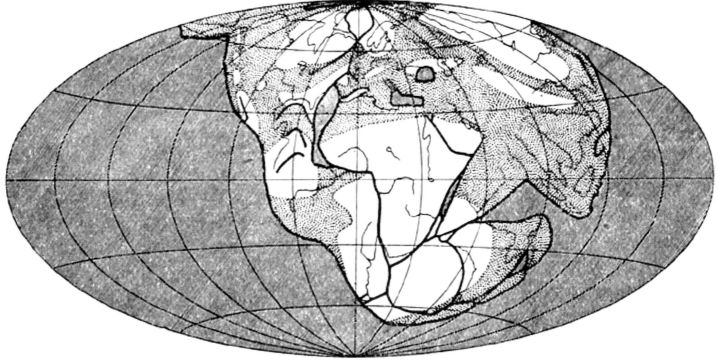

there were long-lost land bridges that allowed life to spread across the world.

The idea would not go away. If rocks can change and move in a geological formation, then why not continents? Bigger movements might just need more time to take place. Geologists and geographers such as Franklin Coxworthy, Roberto Mantovani and Frank Bursley Taylor all considered notions of continental drift through the late nineteenth and early twentieth centuries. But the scientist most closely associated with the idea, and who coined the term Pangaea, is German polar researcher Alfred Wegener.

A single continent

On 6 January 1912, Wegener unveiled his proposal to the German Geological Society. He argued that the geology along the edges of today's continents indicates that these landmasses were once coalesced into a single supercontinent. Wegener labelled this ancient place Pangaea – roughly, 'all land'. From there, at some later time, Pangaea broke apart and today's continents drifted to where they are now.

The weakness in Wegener's theory was that it lacked a driving mechanism for continental drift. As convincing as the evidence might be that disparate continents were once joined, no one could say what moved them across the Earth. Wegener proposed that the force of the Earth's rotation or changes in the axis of the Earth's tilt could have resulted in all the movement, but other experts argued that these forces wouldn't have been sufficient to do the heavy lifting.

While Wegener's ideas didn't get subducted beneath the geological discourse of the early twentieth century, neither were they widely accepted. Experts who thought that continental drift was a real phenomenon were looked at as being on the fringe, and vanished land bridges were still thought to be the best explanation for why fossil plants and animals from Antarctica to Africa to America were related. Those opposed to continental drift became known as fixists, and they dominated the discourse.

But by the 1950s, geologists had much more evidence than Wegener did in 1912. There were indications that the Earth was made of multiple plates that often joined beneath the ocean. If these continental plates were real, and the Earth's hot, gooey mantle created convection cells beneath them, then the plates could move. All the evidence came together and couldn't be denied. While Snider-Pellegrini, Wegener and others didn't have the entire picture, they saw the pattern that led later experts to the process.

Left The Mid-Atlantic Ridge cuts through Iceland. The Atlantic continues to grow along this divergent plate boundary.

Magnetic striping

Some of the most powerful tools for telling time through Earth's history come from a place that has only recently been explored. Deep beneath the waves of the Atlantic Ocean, along the Mid-Atlantic Ridge, the discovery of magnetic stripes has helped geologists better understand the timing of some of Earth's big events.

Even though people have been travelling over the surface of the world's oceans for thousands of years, it wasn't until the 1950s that we started to get a decent understanding of what the seafloor looked like. Scientists realized that a technique first used during World War II could be applied to learning more about the planet. During the war, naval ships towed magnetometers behind them to detect the presence of hidden submarines. Geologists realized that the same technique could be used to determine the magnetic polarity of seabed rocks below. When the scientists mapped out the polarities of the seabed rocks, they noticed a strange pattern. In some places, such as along the Mid-Atlantic Ridge, they found a zebra-stripe pattern of alternating polarities. The rocks seemed to flip back and forth when it came to where magnetic north was located.

Reversing poles

It wasn't immediately clear what had created the magnetic stripes in the rocks. In the 1960s, a small group of geologists proposed that they were created by the reversal of the Earth's magnetic fields. If you were around during one of these events, your compass directions for north and south would suddenly seem reversed, magnetic north and south seeming to have traded places.

Determining when these reversals had happened required another line of evidence. Thankfully, the striped rocks had been produced from a seafloor ridge where magma reached the surface and 'locked in' the magnetic signature of the planet at the time the rock solidified. These volcanic rocks contained radioactive minerals that decayed according to a constant half-life, and geologists could apply the new science of radiometric dating to the rocks and calculate when these reversals occurred.

Above Volcanic rocks rich in iron, such as this magnetite, can be strongly magnetic.

Opposite The Mid-Atlantic Ridge extends for 40,000 kilometres. At its centre is a rift valley where the continental plates are diverging.

The method of choice for this project was potassium–argon dating. Much like other forms of radiometric dating such as uranium–lead, the technique works because an isotope of potassium decays at a constant rate to an isotope of argon. By looking at how much of each is present inside the ocean rocks' minerals, the experts were able to calculate when those rocks had been laid down and what the magnetic profile of the planet was like at that time.

Charting the reversals

Geologists were able to trace when magnetic reversals had happened back to about 180 million years ago. These times are categorized as either normal – like our present time – or reverse. Examples include the Gauss normal flip, between 3.4 and 2.48 million years ago, and the Matuyama reverse, between 2.48 million years and 730,000 years ago. Back and forth, the rocks recorded reversal after reversal. These rocks can now be compared to volcanic rocks elsewhere. If a paleontologist finds a fossil overlain by volcanic rock, for example, they can compare the polarity of the rock to what is known from the seafloor and come up with an estimate of when that animal was alive.

Spreading seafloor

Dating the seafloor rocks also led to the discovery that the further the rocks were from the ocean ridges, the older they were. This provided critical evidence for seafloor spreading, a process in which new rock is added to the sea bottom, pushing the older rock away like an immense geological conveyor belt. Given that plate tectonics and continental drift were still controversial ideas at the time, this discovery helped provide another line of evidence for the fact that the world's continents had moved.

So far as geologists have been able to tell, these magnetic reversals happen at random. There have been about 183 reversals in the last 83 million years, but they all lasted for different lengths of time. To date, nobody has been able to work out exactly what causes the reversals. We know that the Earth's magnetic field is produced by convection currents of iron in the planet's core, which create electric currents and hence magnetic fields, but no one knows what would make the magnetic poles flip. Dramatic as these events may sound, though, we don't have to worry. There is no evidence that the reversals led to mass extinctions or other disruptions to life on Earth, even though some animals rely on magnetic fields to guide their migrations. We know that these events happen, but the why remains to be discovered.

Above The NOAA's Hercules ROV (Remotely Operated Vehicle) collects rock samples from the Atlantis Fracture Zone on the Mid-Atlantic Ridge.

Opposite Pillow lava on the flanks of a submarine volcano on the Mid-Atlantic Ridge. This rock formation is created when lava cools and solidifies rapidly in contact with cold water.

Lyme Regis

She sells seashells by the seashore. You're probably familiar with this children's tongue-twister. But it doesn't refer to the clam or scallop shells you might find on the beach today. The shells in the rhyme are old, over 180 million years old, and were sold by a pioneering paleontologist who helped to kick-start our fascination with Deep Time. Her name was Mary Anning, and Lyme Regis on England's southern coast was her fossil hunting ground.

You can still find fossils along the beaches of Lyme Regis today. The shores run right up to towering cliffs of crumbling stone that came together early in the Jurassic Period, 200 million years ago. These strata contain the remains of coil-shelled squid relatives called ammonites, fish, shark-shaped reptiles called ichthyosaurs, and plesiosaurs with long necks and four paddles to swim with. Sometimes you can even find terrestrial creatures such as dinosaurs and flying pterosaurs. Anning found all these and more during her fossil forays in the first half of the nineteenth century, and she supplied early paleontologists with fantastic specimens unlike anything anyone had seen before.

Anning was not the first person to walk these shores. Fossil hunting was a business she had learned from her father, and her brother often tagged along with her. She also picked up pointers from her friend Elizabeth Philpot, who was so skilled at her craft that she was able to extract prehistoric ink preserved in cephalopod fossils and use it to draw illustrations with Jurassic pigment. But Mary Anning was undoubtedly the most determined of the fossil hunters combing the Lyme Regis shoreline. Included among her greatest finds were fossils of the fish-like marine reptile *Ichthyosaurus*, the

long-necked and snaggle-toothed *Plesiosaurus*, and the leathery-winged pterosaur *Dimorphodon*.

Back in the Jurassic, the rocks along the Dorset coast were at the bottom of the ocean. Above, the waters teemed with unusual species. Sharks with long spines jutting from their fins, squid

Above A complete fossil specimen of the plesiosaur *Thalassiodracon hawkinsii*, discovered in Somerset, England.

Opposite A letter from Mary Anning in 1823, with a detailed drawing of a plesiosaur skeleton.

Scale One Inch to each Foot

Sir

       I have endeavoured for a rough sketch to give you some idea of what it is like. Sir you understood me right in thinking that I said it was the supposed plesiosaurus, but its remarkable long neck and small head, shows that it does not in the least verifie their congeneros; in its analogy to the Ichthyosaurus, it is large and heavy, but one thing I may venture to assure you it is the first and only one discovered in Europe, Colonel Birch offered one hundred guineas for it unseen, but your letter came one days post before but

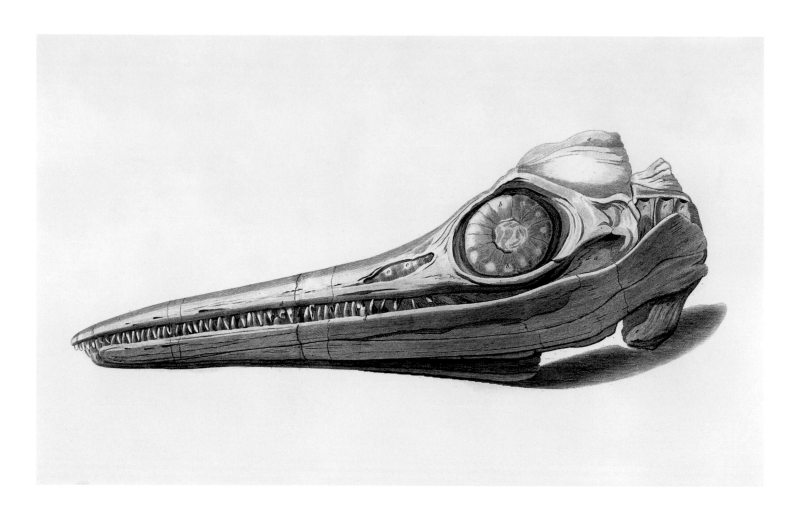

supported by bullet-like internal cones, and a varied menagerie of seagoing reptiles swam these waters. When they perished, they fell to the mucky sediment on the bottom. In different circumstances, scavengers would have picked apart the carcasses and recycled shell, flesh and bone back into the sea. But the conditions on the sea bottom were stagnant, and few scavengers could tolerate life in the ooze. Consequently, many of the bodies from above went undisturbed, buried by fine sediment that delicately preserved so many prehistoric details. Walking along the sand, Anning was looking for the broken remnants of this lost world.

Sidelined by the establishment

An expert through experience rather than academia, Anning took notes and illustrated many of her finds. But at a time when science was deemed unsuitable for women, she was not encouraged or even allowed to write scientific descriptions of these animals herself. That task often went to men in established museum or university positions. Anning struggled with this. "The world has used me so unkindly, I fear it has made me suspicious of everyone," she wrote. Today, she'd be heralded as an exceptional fossil expert, but in her own era, it was male academics who gained renown from her work.

Anning's discoveries among southern England's Jurassic beaches set the stage for a whole new view of the past. The idea that species could go extinct only started to catch on with naturalists in 1799, the year Anning was born. People had been discovering and wondering about fossils for centuries, and indigenous cultures all over the world recognized that fossils represented once-living animals, but nineteenth-century science was just starting to catch up. No one had ever seen creatures quite like ichthyosaurs and plesiosaurs before. No argument could be made that these creatures still lived – someone would have seen them. They clearly lived during an entirely different era, an Age of Reptiles that took place long before our own and then suddenly ended. Without the fossils Anning diligently found, uncovered, cleaned and offered to scientists, this lost world would have remained a mystery for much longer.

Above A scale drawing of a fossil ichthyosaur skull, excavated from the rock by Mary Anning after it had been spotted by her brother Joseph in 1812.

Opposite A large ammonite fossil lying on the beach at Lyme Regis.

The Morrison Formation

If you're looking for dinosaurs, there may be no better place than the Morrison Formation in North America. Jurassic classics including *Stegosaurus*, *Diplodocus*, *Apatosaurus* and *Allosaurus* are all found within these rocks, with new discoveries made every year. The formation represents a rich, vibrant world filled with some of the strangest and largest creatures ever to walk the Earth.

We're fortunate that there's a lot of Morrison Formation to explore. Covering an area of more than 1.5 million square kilometres, the purple, maroon and grey outcrops of this famous slice of Jurassic time can be found from North Dakota to Texas, from Kansas to New Mexico. The rocks were laid down between 156 and 146 million years ago by rivers, lakes and ponds, and in low-lying areas where non-avian dinosaurs roamed vast, fern-covered floodplains dotted with conifer forests.

The Bone Wars

The fossil riches of the Morrison Formation came to the attention of paleontologists early on. During the late nineteenth century, American paleontologists Edward Drinker Cope and Othniel Charles Marsh waged the 'Bone Wars' to discover and name as many fossil animals as they possibly could. The Morrison Formation outcrops of the American West were prime territory, not just for the amount of fossils found there but also for the proximity of the field sites to newly constructed railroad lines. Many of our favourite dinosaurs – including *Brontosaurus* and *Ceratosaurus* – were found as a result of this cantankerous competition. Following Cope and Marsh, the next generation of paleontologists ventured back to the Morrison Formation deposits to find even larger and more complete dinosaurs. This 'Second Jurassic Dinosaur Rush' provided

some of the founding fossils for institutions including the American Museum of Natural History, the Carnegie Museum of Natural History and the Field Museum.

Even after more than a century of research, paleontologists are still finding new sites and species within the Morrison Formation. Some fossils are from absolutely gigantic dinosaurs that rank among the biggest known. The long-necked herbivores *Supersaurus* and *Diplodocus hallorum* are two of the longest dinosaurs of all time, stretching over 30 metres from nose to tail. However, some of the most important recent discoveries have been of smaller creatures. Paleontologists have uncovered the bones of one of the earliest known snakes, *Diablophis*, as well as small, termite-eating mammals like *Fruitafossor*. In fact, the most common fossils in the formation are not dinosaurs at all, but a type of freshwater clam often washed in among the larger boneyards.

Quarries

The Morrison Formation is made up of three separate units, and most of the fossil discoveries are made in the topmost section called the Brushy Basin Member, which is exposed by quarries. These areas became prime destinations for fossil hunters in part because of the sheer number of massive quarries found there, including the Carnegie Quarry at Dinosaur National Monument, the Howe

Quarry in Wyoming, the Cleveland-Lloyd Dinosaur Quarry in Utah, and the Mygatt-Moore Quarry in Colorado. These are all expansive bonebeds that contain multiple dinosaurs of various species, including some dinosaurian rarities. No two Morrison Formation quarries are quite alike, and each has led to new insights. Excavations at the Mygatt-Moore Quarry, for example, revealed an early ankylosaur named *Mymoorapelta*, indicating that ankylosaurs evolved alongside the plate-backed *Stegosaurus* rather than afterwards.

Varied ecosystem

Despite the large numbers of quarries and discoveries, the Morrison Formation still represents a perplexing time and place in Earth's history. This was the Serengeti of the prehistoric world, only on a much larger scale. Each habitat often hosted multiple carnivores over 6 metres long, with multiple species of long-necked

Above A Camarasaurus skeleton at Dinosaur National Monument in Utah.

herbivores that could stretch more than 30 metres from nose to tail tip. Add in all the small mammals, reptiles, amphibians and smaller dinosaurs, and you have an ancient environment that must have been brimming with plant life in order to support so many animals.

If the dinosaurs had lived somewhere else, we might not have known anything about them. Fossils are most often preserved in lowland habitats where sediment is being deposited. (This is why we don't know much about mountain dinosaurs, because they lived in places that were being eroded.) What's more, the Morrison Formation was a highly seasonal habitat. Similar to eastern Africa today, there was a dry season and a wet season. Dinosaurs that died from thirst or lack of food during the dry season would be swept up by the local floods caused by returning monsoons, their bones washed into riverbeds and ponds where they could be quickly buried. Thanks to these happenstances of topography and climate, we have a record of some of the most impressive and fearsome dinosaurs ever to have lived.

Above The skull and jaw of *Dryosaurus altus*, an iguanodont found in the Morrison Formation that lived in the Late Jurassic period.

Left The Bentonite Hills in Utah, composed of the Brushy Basin Member of the Morrison Formation.

The Weald

In building his case for evolution through natural selection in *On the Origin of Species* (1859), Charles Darwin knew that he had to use as many lines of evidence as he could muster, especially from geology. The age of the Earth wasn't known during Darwin's time, and he knew that critics of his evolutionary theory might object that there simply was not enough time for natural selection to change one form of organism into another. Darwin knew that he needed time – Deep Time – so he turned his attention to the Weald in southeast England.

At first glance, the Weald might just look like rolling green countryside. However, a geologist's eye might pick up the parallel slopes of white chalk laid down by a Cretaceous sea and later winnowed down by erosion. The chalk slopes of the North and South Downs are two parts of what was once a dome of rock that had the middle portion broken up. Darwin knew that this kind of landscape must have taken a long time to form, so he set about calculating just how long it would have taken the valley of the Weald to be eroded from its parent rocks. Operating with an estimated rate of erosion, Darwin worked backwards from the present and reasoned that it would have taken about 300 million years for the Weald to form. If that were true, Darwin reasoned, then the world itself must be much older, providing plenty of time for dramatic evolutionary changes to life.

How old is the Earth?

Other experts weren't convinced. In 1862, William Thomson (later Lord Kelvin) calculated that the Sun had been burning for between 20 and 100 million years, declaring that it was 'almost certain' that our star was less than 500 million years old. Darwin wasn't swayed by Kelvin's argument, but he was worried that his own maths might be off, and he removed his geological clock argument from subsequent editions of *On the Origin of Species*.

Of course, we now know that the Earth is very, very much older than 100 million years. About 46 times older than that, in fact. And while Darwin's numbers weren't exactly on point, he was right that the Weald was very old. The sediment that makes up the Weald clays was deposited about 130 million years ago and was pushed up by mountain-building between about 20 and 10 million years ago.

Cretaceous fossils

The significance of the Weald is not just in the age of the rocks. Some of England's – and paleontology's – most significant discoveries were made among the Cretaceous clays of this area. In 1822, for example, Mary Mantell noticed some strange fossils that were turned up as part of a road-building project in the Weald. Among them was a tooth. Mantell's husband, Gideon, had a keen interest in fossils and eventually proposed that the tooth must have belonged to a huge, iguana-like reptile. He coined the name *Iguanodon* for the animal in 1825. Just a few years later, in 1834, multiple bones

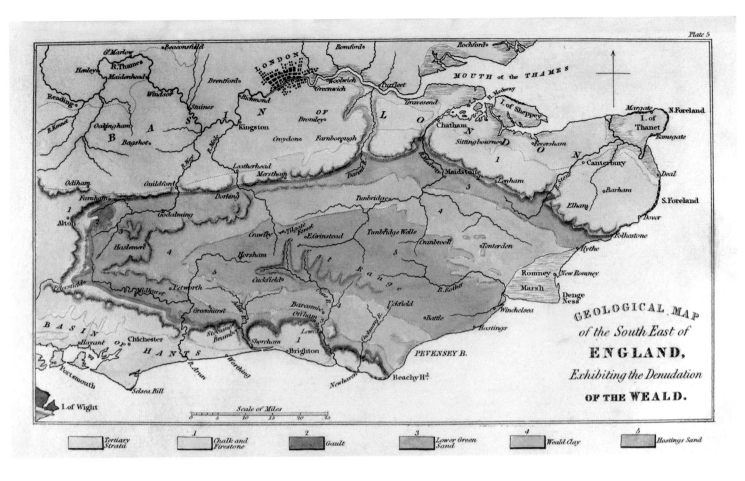

Above A geological map of the Weald produced by
British geologist Charles Lyell in 1833.

Below A speculative drawing of *Iguanodon* by Gideon Mantell.

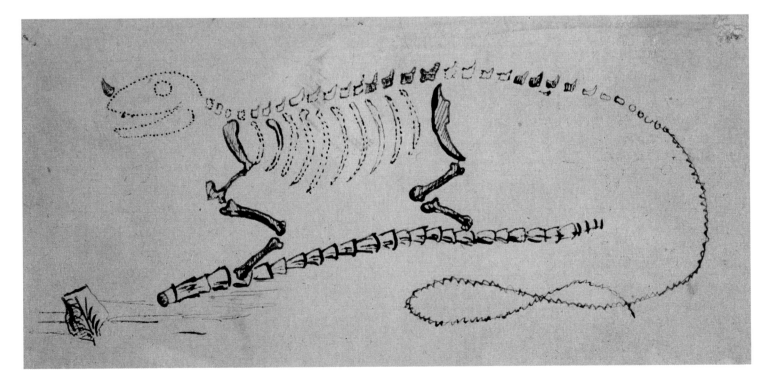

from a single *Iguanodon* were found in a quarry in Maidstone, Kent. This was the first partial skeleton of a dinosaur recognized by science, such a claim to fame that an *Iguanodon* now appears on the Maidstone coat of arms.

Gideon Mantell initially envisioned *Iguanodon* as a lumbering lizard, perhaps 30 metres long if the proportions of modern iguanas were anything to go by. But the Maidstone find and others indicated that *Iguanodon* didn't look anything like a lizard. This was a bipedal reptile with its legs held beneath its body. And the distinctive spike that Mantell thought went on the nose was actually the dinosaur's thumb.

New dinosaur

What *Iguanodon* used its thumb spike for is a mystery. If it was a defensive weapon, though, the dinosaur would have had good reason to keep it handy. In 1983, amateur fossil collector William Walker was poking around the Wealden clay when he found a huge claw. The fossil came to the attention of Natural History Museum scientists, who visited the site and found even more bones. In the end, these fossils turned out to belong to a never-before-seen dinosaur named *Baryonyx walkeri*, or 'Walker's heavy claw'.

Baryonyx was a predatory dinosaur, but very much unlike *Allosaurus* or *Tyrannosaurus rex*. This dinosaur had a long, crocodile-like snout and curved claws on its hands. The fossil was so well-preserved that pieces of fish and a small *Iguanodon* were found in the area of its gut. Paleontologists soon realized that this new animal was related to the great sail-backed *Spinosaurus*, and *Baryonyx* suddenly became the most complete spinosaur known. I wonder what Darwin would have thought of it.

Above A restoration of a corpse of *Baryonyx walkeri* lying on the bed of an ancient lake.

Opposite A section of the South Downs, part of the Weald.

Archaefructus

Out of all the subjects Charles Darwin studied during his career, he dubbed the origin of flowering plants an "abominable mystery". Looking around the world today, it might be hard to imagine that there was a time when trees didn't sport beautiful petals and wildflowers didn't sprout up from fields. But geologically speaking, the fossil record shows that flowering plants are relatively new.

The first photosynthesizing organisms go back billions of years. The earliest plants to grow on land spread into the terrestrial realm about 470 million years ago, long before some vertebrates would make their own transition to land. The earliest gymnosperms – the group of plants that includes pine trees – evolved about 390 million years ago. But flowering plants – technically known as angiosperms – may not have evolved until as recently as 125 million years ago, during the heyday of the non-avian dinosaurs. The critical evidence for this date is an unassuming plant found in the ancient rock of China.

Reproductive parts

Archaefructus doesn't look quite like a rose or a magnolia tree. This small plant appears somewhat wispy, and as far as paleobotanists have been able to discern, it did not have bright, showy flowers. But the presence of certain key characteristics has allowed experts to identify this plant as one of the earliest angiosperms yet found.

Two of the characteristic reproductive parts of flowering parts are the carpels and the stamens. The carpels are part of the gynoecium, or female reproductive parts, while the stamens are male reproductive parts. Both of these parts are only found among angiosperms. In *Archaefructus*, the carpels and stamens are both part of the plant's stem.

To date, paleobotanists have identified three species of *Archaefructus* – *A. liaoningensis*, *A. eoflora* and *A. sinensis*. Some experts see the plant as a representative of what the first flowering plants were like, while others see these plants as related to today's waterlilies. Either way, this ancient plant probably points to even older discoveries that are waiting to be made.

Older origin

Genetic data from modern plants hints that the earliest angiosperms evolved sometime during the Jurassic, predating *Archaefructus* by at least 20 million years. Paleobotanists just haven't found the fossils yet. That might be due to a combination of two factors. First, a case of mistaken identity – the first angiosperms probably didn't look very much like their modern counterparts. And second, the vagaries of fossil preservation – the fossil record is only a fraction of a fraction of a fraction of all life that ever lived, often favouring relatively large or sturdy organisms. Early flowering plants were

Opposite Top A facsimile of a fossilized *Archaefructus liaoningensis*.

Opposite Bottom The relatively simple structures of early flowering plants have diversified into a wide range of forms, such as this beautiful star magnolia.

not only rare compared to other plants, but they were probably also relatively delicate, requiring exceptional circumstances for fossilization. Even *Archaefructus* comes from fossil beds where fine-grained, detailed preservation was common, which is sadly not the case in all fossil deposits.

Whenever angiosperms first evolved, though, these plants set the stage for an array of major evolutionary changes. Prior to the evolution of angiosperms, the most prominent plants on land were conifers, ferns, cycads and similar plants. These plants relied on different ways of reproducing. Ferns reproduced through spores, while cycads and conifers produced cones that sent out pollen to meet cones on other trees. These plants often relied on wind or other aspects of the environment to bring the necessary gametes together. But with their seeds kept inside flowers, or their early forerunners, angiosperms created a new form of reproduction that forever changed life on Earth.

Evolutionary edge

Flowering plants didn't have to wait for a fortuitous breeze to reproduce. Animals – from insects to small mammals – climbed in and around early angiosperms, helping to pollinate them. In fact,

paleontologists have been lucky enough to find direct evidence of ancient pollination in the fossil record. In a piece of roughly 99 million-year-old amber from Myanmar, paleontologists have identified a Cretaceous beetle that carried pollen grains on its body. The beetle even seemed to have a body shape and mouthparts similar to those of pollen-feeding beetles today, indicating that these interactions had already been underway for a very long time.

This new way of pollination gave angiosperms an edge. Plants with an attractive colour, scent or fruit could out-reproduce other forms. Over time, angiosperms began to outstrip conifers in some ancient habitats as the primary forms of plants. During the Cretaceous, flowers started to take on their vivid splashes of colour and attractive shapes. By the time of *Tyrannosaurus* and *Triceratops*, 66 million years ago, plants such as dogwood trees were already blooming. These dinosaurs could have stopped to smell the flowers if they had wished to.

Above Typical flora from the Early Cretaceous, before flowering plants.

Opposite A dorsal view of *Cretoparacucujus cycadophilus*, a beetle trapped in amber with grains of pollen 99 million years ago. Its mandible appears to be specially adapted to collect pollen.

Metasequoia

Imagine walking through a forest, looking at all the wonderful forms of life around you, and having a *Triceratops* cross your path. This probably sounds like something out of a time travel novel or other form of science fiction, but it isn't too far off from what happened with the discovery of the dawn redwood tree, or *Metasequoia*.

In July 1943, Chinese forestry official Zhan Wang heard a rumour that he just had to follow up on. In the eastern part of the country, in a town called Moudao, there was a tree that nobody could identify. Wang wasted no time in tracing the rumour back to its source. The tree, a large conifer, was certainly unusual. Wang took some branches and cones as samples to study later.

On closer inspection, the parts of the mystery tree didn't match any known species in China. Wang guessed that the tree might be some form of swamp cypress, but many of the fine details were wrong. The following year, botanist Zhong-Lun Wu happened across Wang's clippings and picked up the thread. Experts agreed that the tree wasn't like any other living plant. And yet botanists had seen this tree before. It was just in stone instead of a forest. What Wang had found was a specimen of *Metasequoia*, or dawn redwood, a genus of tree that had been named in 1939 from fossils tens of millions of years old. Four years separated the discovery of this tree as a fossil and finding one alive.

Lazarus species

The dawn redwood was the dendrological equivalent of the coelacanth. This was a tree that was recognized first from fossils, and presumed extinct, before turning up alive and well in our modern world. Today, such species are known as Lazarus taxa, after the New Testament fable about a man raised from the dead.

In the case of *Metasequoia*, this tree has been around for about 100 million years. Its fossils are found across the Northern Hemisphere, split into three separate species. These were no bit players in life's drama, either. Around 55 million years ago, the world's climate experienced a rapid heat spike. Temperatures rose by more than five degrees Celsius in a short amount of time, melting ice at the poles and allowing life to spread to places that were otherwise inaccessible. On the ancient Ellesmere Island, within the prehistoric Arctic Circle, great forests of *Metasequoia* grew near the North Pole, while alligators slid through the swamps and early primates clambered through the trees.

Below A 49-million-year-old fossil of a *Metasequoia occidentalis* branchlet.

Opposite A living dawn redwood, species *Metasequoia glyptostroboides*.

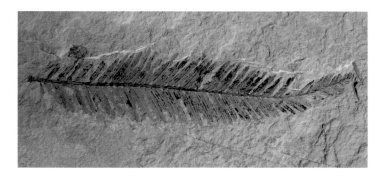

Living clues to the past

Thanks to living *Metasequoia*, paleobotanists have been given some deep insights into what these prehistoric forests were like. The dawn redwoods had thick, rough bark, much like their sequoia namesakes that grow in the Pacific Northwest of the United States. Dendrologists have also discerned that *Metasequoia* grow fast, reaching about 45 metres tall and 2 metres around, and like other conifers they reproduce via cones. Unlike classic conifers, though, dawn redwoods are deciduous trees that shed their leaves during winter seasons. Paleobotanists hypothesize that this might have been an adaptation to life at high latitudes during the ancient past, when prolonged months of darkness would have caused dawn redwoods to temporarily lose their foliage.

Missing records

Despite the fact that dawn redwoods have been around since the Late Cretaceous, their fossil record is uneven. Although we now know that *Metasequoia* survived, the plant's fossil record vanishes during the Miocene, or between 23 and 5 million years ago. Exactly why has left paleobotanists stumped. In the case of the coelacanth, it seems the fish evolved to survive in deep water habitats that are rarely fossilized. But *Metasequoia* live in terrestrial habitats and have an extensive fossil record in certain places and times.

This conundrum is an illustration of how uneven the fossil record can be. If a species becomes relatively rare, it may seem to vanish from the fossil record. Then again, even a common species has to live in a place where sediment is actively being laid down to have any hope of fossilization. *Metasequoia* that may have grown in mountain habitats, which eroded down over time and are practically invisible to us today. Or perhaps something about the tree's biology shifted, causing it to live in places that are not as favourable to preservation. We know that the tree survived and has persisted for far longer than any of the dinosaurs that once walked beneath its branches, but the fossil record still keeps the secret of its exact means of survival.

Above *Metasequoia* leaves in autumn, ready to shed.

Opposite A dawn redwood forest in Zhejiang Province, China.

Chicxulub Crater

One of the Earth's biggest and most painful scars is hard to see, despite the fact that it is awfully large. The edge of the 150 kilometre-wide crater bites into the Yucatan Peninsula in southern Mexico, while the rest sits 20 kilometres beneath the waves in the Gulf of Mexico. Driving at motorway speeds, it would take over an hour to get from one side to the other, while its depth is enough to conceal some mountains. This depression, known as the Chicxulub Crater, is the remnant of what was arguably the worst day in the history of life on Earth.

About 66 million years ago, on an ordinary Cretaceous day, a chunk of rock over 11 kilometres wide slammed into the Earth. The bolide was too fast to even see as it entered Earth's atmosphere and buried itself into the crust. Moving at a speed of about 20 kilometres per second, the asteroid or other piece of extraterrestrial debris gave the planet a horrible wallop that triggered Earth's fifth mass extinction.

The devastation began immediately. At the site of impact, heated rock and debris was shot back out into the atmosphere, while immense tidal waves were sent out from the centre of impact. The waves rebounded on the coasts and crashed back into the crater. Creatures in the immediate area – from the mightiest dinosaurs to the smallest plankton – were wiped out in a flash.

If that had been all, maybe life elsewhere would have continued unaffected. Mammals would have stayed small, dinosaurs would have continued to dominate the land, and evolution would have taken a very different path in a continued Age of Reptiles. But that's not what happened. Within hours, the pulverized chunks of rock that were sent up into the atmosphere rained down at places all around the planet. Some of them sparked raging forest fires that spewed smoke and soot into the air. Worst of all, the friction of

Above The impactor devastated the surrounding area.

Opposite A core sample of material taken from the impact crater. The lighter areas are fragments of rock, mostly anhydrites and evaporites.

falling heated the atmosphere to oven-like temperatures. Many animals that were too large to hide in a burrow, slip into the water, or otherwise hide from the heat died on that first day after the impact.

Three years of winter

Even then, much life might have been spared. But the Chicxulub impactor hit a particularly unfortunate place on the planet. Not only did all the dust and rock and soot begin to block out sunlight soon after impact, but the rock that the impactor hit was itself rich in sulphate minerals. Strewn through the air, the sulphate aerosols began to cool the global climate and triggered an impact winter that lasted at least three years. Photosynthesis totally stopped. On land and in the seas, no new plants grew. Surviving creatures either had to eat the sea itself or each other to persist through the long dark, and most did not last long. Paleontologists estimate that about 70 per cent of all known species perished from either the intense heat or the dark cold that followed it.

Almost all the dinosaurs died out, leaving only birds to carry on their legacy. Also wiped out were the flying pterosaurs, the sea-going mosasaurs, and other forms of reptiles that had thrived for millions of years. Coil-shelled squid relatives called ammonites, toilet-seat-

size clams called rudists, and other invertebrates died, too, and even among the surviving groups there were mass extinctions of mammals, lizards and birds. The oceans were so badly affected that they almost went back to a soup of single-celled organisms that hadn't been seen in a billion years. Life in every habitat was affected by the impactor's strike in what is now known as the Cretaceous–Paleogene (K–Pg) extinction event.

This emerging picture of hell on Earth has only recently started to come together. Experts only identified asteroid impact as a possible extinction cause in 1980, based on a spike in an element called iridium. This metal is rare on Earth, but common in asteroids, and is abundant in the rock layer right where the likes of *Tyrannosaurus* disappear in the fossil record. Experts hotly debated the issue; no one had ever found evidence of an impact-caused extinction before. What they didn't know was that the Chicxulub Crater had already been discovered. Oil geologists Glen Penfield and Antonio Camargo found the first signs of the crater in 1978, and in 1990 all the pieces came together. The iridium-rich rock recorded the impact's global effects, and the crater itself was the smoking gun. This place, this one crater, represents a single moment that altered the course of history forever.

Above An artist's impression of the Chicxulub impact.

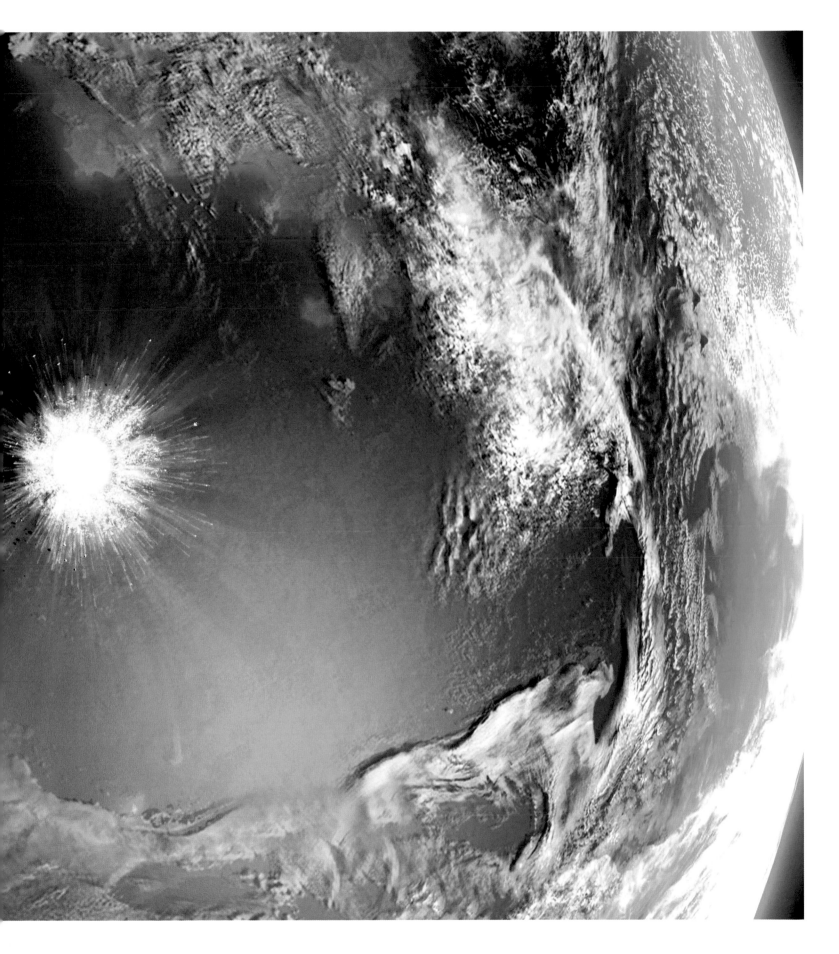

Building mountains

There is hardly a geological phenomenon more impressive than a mountain. Whether covered in snow or forest, dotted with rocky crags or pleasant enough to hike up, mountains often lead us to wonder how such towering and seemingly permanent parts of the landscape got there in the first place.

The technical term for mountain-building is 'orogeny'. It has happened many times throughout the course of Deep Time. Entire ranges have risen, been eroded back down and disappeared, then their rocky leftovers used to form the basis of new mountains. Even mighty Mount Everest is not permanent, but an ephemeral part of our world – the Himalayas were pushed up around 50 million years ago, when the island subcontinent of India collided with Asia.

Mapping the Alps

Much like many other geological phenomena that we now take for granted, the puzzle as to how mountains formed fostered a great deal of debate. One of the key players in this particular story was French geologist Marcel-Alexandre Bertrand. Born in 1847, Bertrand was the son of mathematician Joseph Louis François Bertrand and he, too, became a scientist. Rather than pursuing maths, however, Bertrand focused on geology. After training at the École des Mines de Paris, Bertrand focused on mapping the Jura Mountains – the origin of the word 'Jurassic' – and the Alps.

Bertrand was fascinated by the history of mountains. The idea that mountains were made of rock laid down in other environments – such as seabeds accumulating sediment, or volcanoes oozing lava – was well accepted by his time. But that only explained where the parent rock came from. The forces involved in mobilizing those rocks into mountains were unknown. Faced with this conundrum, in 1887, Bertrand proposed an idea that was well ahead of its time.

Dynamic origin

From his research on the Alps, Bertrand proposed that mountains were created by motions of the Earth's crust itself. Rather than being static, the rocks of the Earth's crust could change shape and move. Imagine, then, that one area of the Earth's crust started to become squeezed by surrounding rock. The part under pressure would grow thicker and horizontally shorter, just like trying a rubber ball being squeezed at the sides.

But there's only so much squeezing that rock can take. Eventually, Bertrand suggested, the tense part of the Earth's crust might pop back up and low-lying rocks might be thrust into mountain ranges. Today, geologists know this phenomenon as overthrusting.

Bertrand didn't think that this was a one-off or especially rare event. In fact, his vision of great folds in the Earth's crust and resulting thrusts had created many of Europe's mountain ranges. In his view, there were Caledonian, Hercynian and Alpine orogenies, creating some of the most impressive topography on the continent. This idea came to be known as the wave concept of mountain building, after the idea that folds and thrusts came in waves over time.

Opposite The Jura Mountains studied by Bertrand started to form 65 mya.

Crumpled rock

Part of Bertrand's hypothesis involved a geological structure called a nappe. This is a large sheet of rock that has been moved more than a kilometre above its original position due to extreme compression. The phenomenon's name is a clue to how it looks to a geologist's eye. The word 'nappe' comes from the French word for tablecloth, and a rocky nappe somewhat resembles a crumpled tablecloth sliding across the smooth surface of a table.

These geological curiosities are still studied by experts today. Bertrand's breakthrough came from not merely documenting them but also asking deeper questions about them. "The idea of making a fault a subject of study and not an object to be merely determined," he wrote late in his career, "has been the most important step in the course of my methods of observation. If I have obtained some new results it is to this that I owe it."

This picture of an active, changing Earth was different from that which many other nineteenth-century geologists expected. The idea that geological forces in action today could explain past events – Hutton's concept of uniformitarianism – was only decades old. And even then, these forces seemed relatively minor compared to the broad-scale geological events that could cause continents to move or mountains to rise. Bertrand's idea, after all, required a sense of tectonics, an idea whose time was yet to come. For there to be folds and thrusts to create mountains, the Earth's crust had to move and even change shape. Geological consensus wouldn't arrive at this point until about six decades after Bertrand's death in 1907.

Above Marcel-Alexandre Bertrand (1847–1907).

Right Chief Mountain, Montana, is part of the Lewis Overthrust, a 300-kilometre-long overthrust fault in North America.

Green River Formation

Most of the time, the fossil record comes to us in bits and pieces. Skeletons are incomplete, bonebeds are jumbled, or experts find animals but not plants, each part telling us something new but still as perplexing as an individual jigsaw piece split from the rest of the puzzle. But there are some places where paleontologists get a look at the vast sweep of an ancient ecosystem. The Green River Formation in the western United States is one such place.

Great fossil slabs from the Green River Formation record what life was like around 50 million years ago. This expanse of Deep Time was laid down during the Eocene, or the 'dawn of the recent'. Back then, in the millions of years after the K–Pg mass extinction (see page 144), the Rocky Mountains continued to rise through western North America. The Wind River Mountains, Sawatch Range, Uncompahgre Plateau, Wasatch Mountains and more all rose and altered the topography of the ancient west. And as mountains went up, low basins between them filled with water across what would one day become Wyoming, Colorado and Utah. These lowlands hosted immense lakes that could span more than 38,000 square kilometres. These were ideal settings to preserve the life in and around the water between 53 and 48 million years ago.

You may have seen Green River Formation fossils before. This geologic formation is rife with small, well-preserved fish, leaves and invertebrates. Fossil collectors and dealers often sell them at truck stops. But service station trinkets don't really capture the full picture that scientific explorations of these ancient lakes have uncovered. When all the leaves, fish, reptiles, mammals, birds, insects and other organisms are viewed as a whole, this formation offers a rare window into a distant time.

Warmer world

Back in the time of the Green River Formation, the world was warmer than it is today. The average annual temperature in the Eocene was between 15 and 20 degrees Celsius compared to a global average of 14.5 degrees Celsius today. Greenhouse gases released in the aftermath of the K–Pg extinction helped to heat this warmer world, where subtropical forests extended over much of the Northern Hemisphere. The fact that fossil alligators and crocodiles are found in these rocks, for example, underscores just how different these parts of the west were during the Eocene.

These vast lakes and bordering forests hosted species that likely would have seemed both strange and familiar. Palms, cat tails, sycamores and willows grew along the lakes, perhaps giving the environments a modern feel. Familiar insects such as assassin bugs, crickets and weevils thrived in the lush environment. But many of the vertebrates that lived here were very different.

Opposite A fossil of a soft-shelled turtle (*Trionychid*) from the Green River Formation.

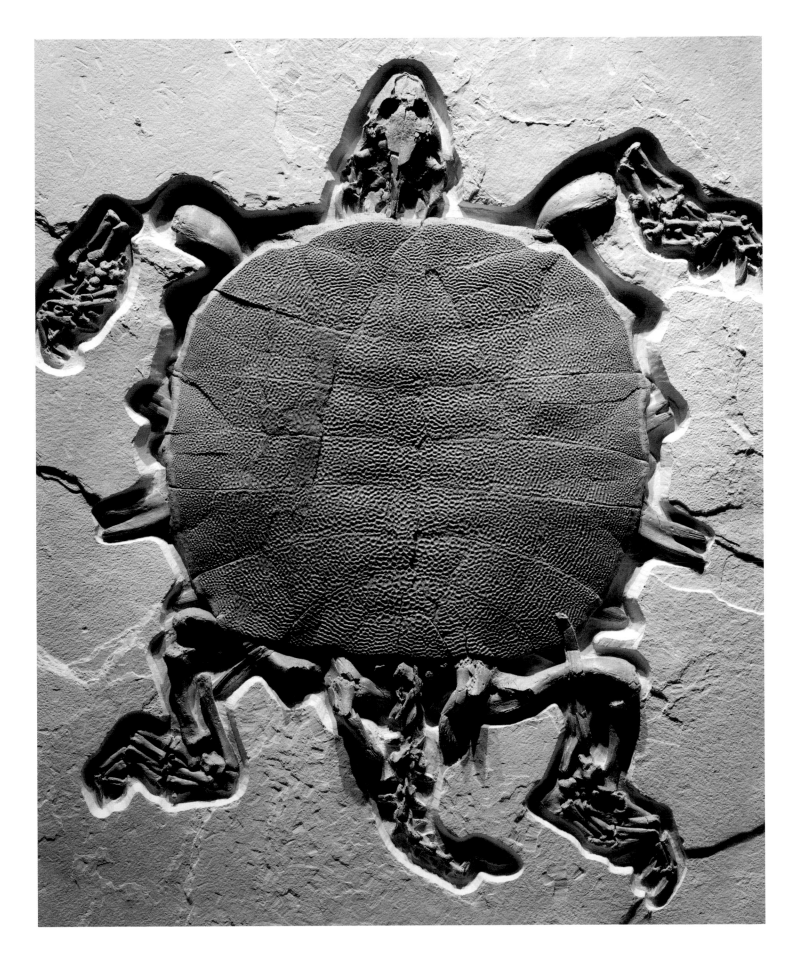

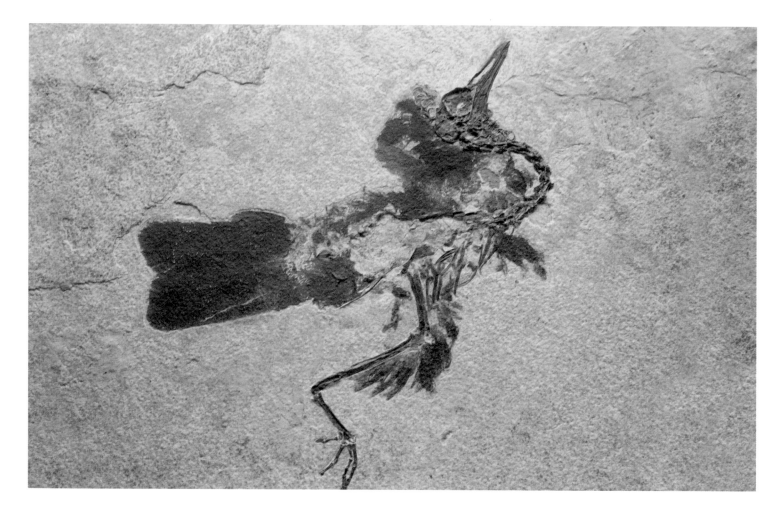

Decades of splitting the tan-coloured slabs of the Green River Formation have yielded two of the oldest known bats, *Icaronycteris* and *Onychonycteris*, which could certainly fly but could not echolocate. Early horse relatives have been found here, too – small, herbivorous mammals about the size of a small dog, with multiple hoofed toes. There were even early, lemur-like primates such as *Notharctus*, an ancient cousin of species that now live only on Madagascar.

Diversifying birds

Some of the most spectacular finds have been of articulated bird skeletons, sometimes found with feathers still surrounding them. The Green River Formation preserved some of the oldest representatives of modern bird groups such as parrots, mousebirds and frigate birds. These individual finds add up to a stunning picture of prehistoric lakes absolutely brimming with life. Altogether, the finds show that ecosystems had fully recovered by the Eocene, with the Age of Mammals in full swing as birds continued the dinosaurian legacy.

The various organisms that called these lakes home didn't become fossilized on land or along the lakeshore. What paleontologists pick away at are the layers of an Eocene lake bottom. Some animals perished above the lake water and fell in. Others died along the lake margins and were washed in by storms or local flooding. Others lived in the lake itself and descended to the bottom.

The bottom waters of these Eocene lakes were oxygen-depleted. These anoxic conditions meant that few organisms lived along the bottom, decreasing the likelihood that scavengers might come along and eat the carcasses. The incredible quality of the preservation – from in-place crocodile scutes (scales) to spider spinnerets – is due to the fact that these various plants and animals were rapidly buried in fine-grained mud. This kept the bodies together. The ancient mud also allowed for a higher preservation fidelity than coarser sediments like sand. It's this exceptional preservation that keeps paleontologists going back year after year, splitting open slabs of shale to see what new vision of Eocene life might come into view.

Above A well-fossilized bird from the Green River Formation in Wyoming. It has been identified as a new species and named *Nahmavis grandei*.

Opposite Fossil Butte National Monument, Wyoming, a rich source of Eocene fossils.

The Martian meteorite

It is difficult to find a more controversial piece of stone than the Allan Hills Meteorite. This piece of Mars somehow found its way to the bottom of the world.

On 27 December 1984, a scientific team searching Antarctica for meteorites found just what they were looking for. Among the continent's Allan Hills, they discovered a nearly 2-kilogram piece of Mars rock. More than a decade later, in 1996, researchers announced that they had found something startling on the meteorite, which is officially known as ALH84001. The rock contained strange tubes and divots that looked like microscopic fossils of bacteria or other microorganisms. Headlines immediately lit up about the potential discovery of life beyond Earth.

Life on Mars?

It's not just David Bowie who has wondered about life on Mars. Scientists have been considering the possibility for a very long time. In 1877, for example, the Italian astronomer Giovanni Schiaparelli noted the presence of lines on the Martian surface that he called *canali*, or channels. When Shiaparelli looked through his telescope at Mars, and the air was still clear enough to get a good look at the planet, he thought that he saw what looked like straight-line divots in the planet's surface, which he then drew. Other astronomers saw the same thing, and even noted the appearance of new channels where there hadn't been any before. The question was what these *canali* really were.

American astronomer Percival Lowell thought he had the answer. These structures had to be canals or irrigation ditches dug by Martians. The appearance of new ditches, as would follow growing seasons, was his explanation for the new lines. Even as other scientists doubted the theory, Lowell expanded his ideas in

a number of books, including *Mars as the Abode of Life*, published in 1908. It wasn't until decades later, as researchers began to invent new ways to look at and even photograph the red planet, that the truth became clear. The canals were not real features of the Martian surface. The lines were optical illusions created by looking for hours through telescopes, and did not exist on Mars itself.

The idea that Mars might harbour some form of life remained tantalizing. By solar system standards, Mars isn't very far away from Earth. It's another rocky planet that looks like it might have once had liquid water, thought to be an important prerequisite for life. By

Above Meteorite ALH84001. About 80 per cent of its surface is covered in a dark fusion crust.

Opposite A true-colour image of Mars taken by the ESA's Rosetta spacecraft during a flyby of the planet in 2007.

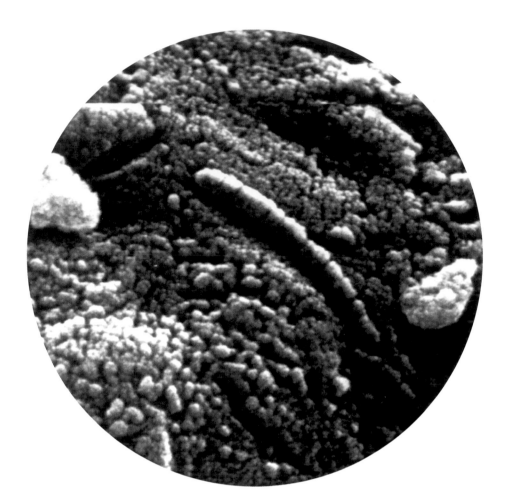

1963, astronomers had discovered water vapour on Mars, perhaps a sign that there had once been life on this planet.

Shot in the dark

Having been discovered long before Mars rovers such as *Curiosity* and *Opportunity* set out to look for possible signs of life on Mars, the Allan Hills Meteorite held the exciting possibility that single-celled life inhabited our nearest planetary neighbour. From the available geologic details, researchers proposed that ALH84001 was a piece of Mars' surface that broke off when another meteorite hit the desert planet around 17 million years ago. The stone then drifted through space until about 13,000 years ago, when it fell to Earth and hit Antarctica. This, in itself, was amazing – that a piece of Mars could be blasted off and somehow make the hop to Earth, only to be discovered millions of years after it left its home planet. That's a distance of over 209 million kilometres – one heck of a lucky shot.

Despite its remarkable life story, what made the meteorite famous were microscopic threads that looked similar to bacteria, but at a much smaller size. Controversy followed. Some experts proposed that the same kind of shapes could be recreated in the lab and that these potential traces were not fossils at all. Researchers who proposed the Martian bacteria hypothesis countered that additional 'biomorphs' have been found in other parts of the Allan Hills Meteorite and in two more Martian meteorites.

Lack of corroborating evidence

The sticking point is that there is no other evidence as to what these structures might be other than their shape. The requirements for microfossils to be recognized as the remains of ancient life are stringent, and they include additional clues aside from just shape. While no one can quite agree on what the tiny structures *are* – be they very minuscule bacteria or mineral growths – the current consensus is that ALH84001 doesn't contain evidence of fossils. Instead, this traveller from another planet comes to us as a mute stone that can only give up clues through scientific questioning. Whether it contains evidence of ancient life or not, the fact that part of Mars could wind up on Earth is amazing enough by itself.

Above The tiny structures on the ALH84001 meteorite are revealed under an electron microscope.

Opposite Geologist Erik Gulbranson working in the Allan Hills.

Oldowan tools

Humans aren't the only animals to make or use tools, but we're certainly the most reliant upon them. Whether we're talking about a hammer, a lever, a screw or a wheel and axle, our whole word is shaped by our ability to create and use tools to literally reshape our environment. And hominins have been doing so for millions of years.

We may never know what the very first tools were like. If modern chimpanzees and other primates are any indication, the first human tools were probably sticks, stones and other slightly modified objects that had little chance of being preserved or even of being recognized unless found in the grip of an early hominin. But we can still be sure that early humans had been inventing and using tools long before the emergence of our own species, *Homo sapiens*.

Much of what we know about early human evolution comes from eastern Africa, and one of the most thrilling anthropological hotspots is Ethiopia's Afar Basin. There, in strata dated to 2.6 million years old, is a site that archaeologists have dubbed Dora 1. At this place in Ethiopia's rocky desert, anthropologists have found a collection of 327 stone tools just a few kilometres away from the place where the oldest *Homo* fossil yet known was found. Even though the jawbone and stone tools were preserved about 200,000 years apart in time, the geographical proximity has led researchers to suggest that an early member of our own genus made these tools rather than another hominin like *Australopithecus* or *Paranthropus*.

Simple technology

As stone tools go, the tools found at Dora 1 are very rudimentary. Many have a smooth edge of unaltered stone with sharper broken edges. The shape makes sense for tools that could easily be created and held in the palms of our hands. At least one smooth edge would have been comfortable to hold in early human hands, while another stone was used to break off flakes to get the desired sharp edge. With a few strikes, an unassuming stone could be turned into a tool.

Anthropologists categorize tools like this as part of the Oldowan Complex. These tools were not all made by the same population or even the same species of hominin, but instead represent an early variety of stone tool that was likely invented and re-invented multiple times through early human history. The moniker for tools like these comes from Tanzania's Oldupai (or Olduvai) Gorge where this type of tool was initially found, but the Dora 1 tools are several hundred thousand years older than those found further to the south.

Opportunist scavengers

Why did humans invent these tools? Bones found at Dora 1 offer tantalizing clues. Along with the stone tools, anthropologists found the bones of herbivores such as gazelles and giraffes at the Dora 1 site. Perhaps the humans who made the Oldowan tools left at the site created them to butcher carcasses that they happened across.

Opposite Top A stone tool found at Dora 1.

Opposite Bottom The earliest stone tools were possibly made to butcher animal carcasses.

Cut marks on bones would seal the case that these were butchery tools. In fact, anthropologists have suggested that primates were using some sort of rudimentary tools to butcher animal carcasses more than 3 million years ago, before the evolution of the genus *Homo*. The Oldowan tools at the Dora 1 site would represent an improvement in this kind of technology, allowing those who made them to cut muscle from skeletons and even crack open long bones to get the nutritious marrow inside.

These humans were not hunters, but rather were opportunists. At 2.6 million years ago, adult humans were 1–1.5 metres tall. They were small, could not run very fast, and had no sharp teeth or claws to defend themselves. And we know from the fossil record of early humans that our predecessors and relatives were often prey to eagles, crocodiles, leopards and other carnivores – from giant hyenas to sabre-toothed cats – that lived in Africa at the time.

Growing brains and changing diets

Putting the pieces together, humans probably started including more meat in their diets as chance opportunities presented themselves. For example, if a sabre-toothed cat like *Megantereon* took down a giraffe, a small band of early *Homo* likely watched and waited from a safe distance until the cat ate its fill. When an opportunity presented itself, the humans moved in to cut and scrape what they could from the carcass and add a little animal protein to their diets. Over time, however, the energetic demands of larger brains demanded more fats and proteins and required that humans improve their technology to hunt food for themselves – or at least be able to drive competing carnivorans off their kills. The advent of tools led to our own reinvention.

Above A fossilized skull of *Homo habilis*, a species of early human that lived between 2.3 and 1.7 million years ago. This is the oldest reliably dated *Homo* species.

Opposite A reconstruction from a skeleton of the hominin *Australopithecus afarensis*. Dated to 3.2 million years ago, the specimen was dubbed 'Lucy'.

Oldupai Gorge

Of all the fossil localities that preserve portions of the human story, none is as famous as Oldupai Gorge. Situated in the northern part of Tanzania along East Africa's Great Rift Valley, this 48-kilometre-long ravine has yielded some of the most informative finds of our fossil past.

The name Oldupai (or Olduvai) comes from the Maasai word for 'the place of wild sisal' after the abundant sisal that grows throughout the area. Beneath that modern overlay is bed after bed of prehistory going back over 1.9 million years. Fossil crocodiles, boneyards created by prehistoric lions, stone tools and of course the bones of early hominins such as *Paranthropus boisei* and *Homo habilis* have been found here.

First discoveries

While Oldupai Gorge has long been important to East Africa's Maasai people, the area's scientific story only started just over a century ago. In 1911, German physician Wilhelm Kattwinkel visited Oldupai Gorge and found the fossil bones of a prehistoric horse with three toes. These finds caught the attention of other naturalists, which eventually piqued the interest of British anthropologist Louis Leakey. Oldupai seemed just the place to find evidence of early stone tools and perhaps the creatures that made them. Over the ensuing decades, Leakey, his family and their collaborators set up a landmark research programme centred around searching the beds of Oldupai for clues about early humans.

Each geologic bed at Oldupai represents a different time period and different species of early human. Bed I, for example, dates to between 1.9 to 1.75 million years ago and contains multiple sites where ancient humans dwelled. Some of the finds that really put Oldupai on the map for anthropologists, however, date to around 2 million years ago. In addition to stone tools, anthropologists have found the remains of the deep-jawed *Paranthropus boisei*, the 'handy human' *Homo habilis* and *Homo erectus*.

The fossil humans in these layers are not all separated in time. They do not represent a smooth gradient from one leading to the next. They are all different forms of early human, and some overlap in time with others. *Homo habilis*, one of the earliest representatives of our own genus, lived alongside *Paranthropus boisei* and the two almost certainly encountered each other. What those encounters were like, however, is anyone's guess.

Dangerous waters

We know a little bit about the hazards early humans faced in this area. Around 1.84 million years ago, what is now Oldupai hosted an ancient lake. In that lake lived 5-metre-long crocodiles with

Opposite A view of Oldupai Gorge looking towards the Naibor Soit Hills. The Bed III site is seen in the foreground.

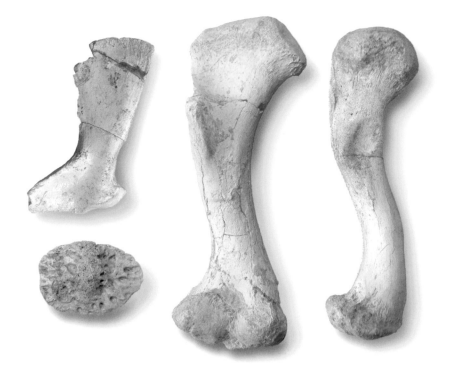

Above *Crocodylus anthropophagus* bones discovered at Oldupai Gorge.

Below An acheulean hand axe, typical of the cutting tools made by early humans.

Opposite Mary Leakey (1913–1996) at work in Oldupai Gorge with her husband
and fellow anthropaleontologist Louis Leakey (1903–1972).

horn-like projections on their skulls called *Crocodylus anthropophagus*. The reptile got its name ('human-eating') because some of the hominin fossils found at Oldupai have crocodile bite marks on them. No one knows whether the crocodile caught unwary humans at the water's edge or was scavenging, but, especially during a time when early humans were relatively small and short-statured, these reptiles were a risk when hominins went for a drink of water.

Early humans almost certainly spent some time scavenging, too. Some of the sites in Oldupai, such as the place where anthropologists found the skull of *Paranthropus* known as the 'Dear Boy', are littered with flakes and debris left over from tool-making. Exactly who made these tools isn't entirely clear. The hunch has often been that *Homo habilis* made these tools, while *Paranthropus* relied on more rudimentary tools made of bone or sticks. Whoever created the tools, though, they were clearly using them to strip flesh from animal bones and get at the marrow inside. Bones of fish, birds and large mammals like wildebeest have been found in the same areas as the tool debris, hinting that the tools were specifically manufactured to help early humans get a little more protein into their diets. It's

unlikely that these humans were hunting – there are no finely crafted spear points or hand axes here – but they would take whatever meat they could get.

Changing landscape

Although the discoveries that made Oldupai famous played out during the middle of the twentieth century, research still continues along the exposed rock of the gorge. Anthropologists have recently excavated a 2–1.8 million-year-old site called Ewass Oldupa, or 'the way to the Gorge' in the Maa language. The locality records changing conditions over the course of 200,000 years, detailed layers representing how this area went from lakeside to fern meadows to woodland to steppe, with ancient hominins coming and going through time.

The emerging picture is that Oldupai was not a single crucible for human evolution. Rather, it represents a varied landscape that rapidly changed over time. What this place meant to people is told in what they left behind, from the tools they made to their very bones. The layers of Oldupai are like a flipbook of ancient human history, every bed another page in the book.

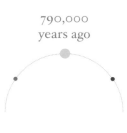

Tektites

About 790,000 years ago, a kilometre-sized chunk of rock fell through the atmosphere and slammed into what is now southern Laos. The critical evidence for this event did not come from an ancient pockmark in the Earth's surface. Instead, the crucial clues are small, pebble-sized globs of glass called tektites.

Some of the Earth's greatest impacts are hidden. That might seem strange given some of the enormous craters visible in our planet's crust today, like the Chicxulub Crater around the Yucatan Peninsula or Upheaval Dome in the Utah desert. But the Earth's surface is always changing, with geological and biological forces constantly reshaping the terrestrial realm. That's why some of the best evidence for significant impacts in the past comes from hints that can't necessarily be spotted with the naked eye.

Throwing glass

Tektites are often hallmarks of impacts on our planet. When a large asteroid, comet or meteorite strikes the Earth's crust, the intense pressure and heat melts some of the sediment at the impact point. This happens incredibly quickly, and the force of the impact ejects tektites back out into the atmosphere as debris. They are often found hundreds or even thousands of kilometres from the site of the impact, creating a halo of ejecta that geologists can track to figure out where major impacts have occurred.

In the case of this particular Ice Age impact, geologists have found tektites from the same time at locations in Australia, Asia and Antarctica. About 20 per cent of the Eastern Hemisphere seems to be littered with them. Something must have struck this part of the world during the Late Pleistocene. The question was where

the bolide struck. Tracking tektites helped researchers zero in on a possible location in southern Laos.

Tektites aren't all the same size; some are larger than others. And the larger a tektite is, the shorter the distance it is likely to travel before coming back down to Earth. By picking up the trail of larger tektites, geologists were able to narrow down the area where a crater might be located.

Above Blocky tektites like this one travel shorter distances.

Opposite The Bolaven Plateau, the possible impact site in Laos, is covered in thick jungle.

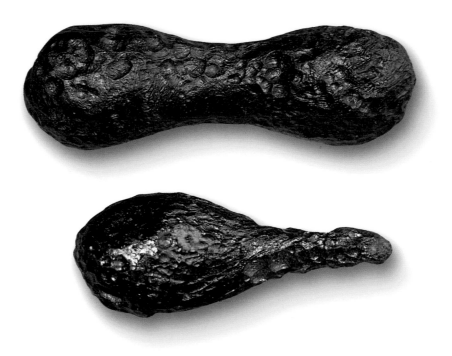

Hidden crater

The best candidate for the impact site so far is on the Bolaven Plateau in Laos. Two intertwined lines of evidence suggest that this is the right spot. The first is that the tektites found in this area have asymmetrical, blocky shapes. They wouldn't have been very aerodynamic and therefore couldn't have travelled very far from the place of impact. On top of that, geologists found a large swathe of hardened lava from an eruption that occurred after the tektites fell to Earth. The extent of the lava flow would have covered up whatever crater there might be in the area.

That's not all. Large impacts sometimes trigger tectonic activity. The impact could have been so disruptive that it helped trigger the eruption that ended up concealing the crater. The fact that there is a gravity anomaly where the suspected crater rests – perhaps caused by a divot filled with thick, dense lava – lends support to this idea. Experts estimate that the crater beneath the lava might be more than 12 kilometres across, with a rim about 100 metres tall.

Averting a crisis

Despite the volatility of this impact on a local scale, no mass extinction or other global biodiversity crisis resulted from the strike. The consequences of an impact hinge on more than simply the size of the bolide. The speed and angle of an asteroid or comet changes how the aftermath plays out, as well as whether the bolide strikes land or sea and the nature of the rock the bolide strikes.

If the asteroid or comet hits rocks that have a great deal of carbon in them, the impact might release that carbon back into the atmosphere and spur global warming. Rocks high in sulphur might have the opposite effect after an impact, causing global cooling.

The impact might have been bad news for ancient hominins living in the area. Humans had already been living in Southeast Asia for tens of thousands of years, and there are localities in southern China where stone tools made by *Homo erectus* are found in the same layers as the tektites. And while the impact did not spark a mass extinction, that doesn't mean that there weren't global consequences. The strike would certainly have thrown a great deal of dust and other debris into the atmosphere, which would have hindered the ability of sunlight to get through Earth's stratosphere. This is similar to what occurred leading up to 1816, dubbed 'The Year Without a Summer' because volcanic eruptions in Indonesia expelled enough dust to dampen sunlight's effects right across our planet. An impact or eruption doesn't have to be massive to have a globally significant effect.

Above Examples of tektites showing two common shapes: the 'dumbell' (top) and the 'teardrop' (bottom).

Opposite Upheaval Dome, a highly visible impact crater in Canyonlands National Park, Utah.

Packrat middens

Some of the best time-keepers in the world are rodents. Not that they know they're doing it. Packrats gather a great deal of material for their desert dens, building comfy homes for themselves and their litters. But given that dry desert environments tend to preserve organic material for long periods of time, combined with packrats' natural propensity for collecting, the nests (or middens) these little mammals create can tell us a great deal about time frames spanning into the tens of thousands of years.

Packrats of America's Southwest go by many names. To scientists, they are species of the genus *Neotoma*. To the public, they might be packrats or wood rats or trade rats. But whatever you call them, they make big, stinky middens. The best spot for a packrat's midden is in a cave, rock crevice, or beneath an overhang. What goes into the den depends on the individual rodent. Plant debris, small rocks, bones, dung and, in more recent times, human litter will all get incorporated into the pile from a radius of about 100 metres around the den. What keeps all this together, though, is not exactly the most charming adhesive. Packrats pee in their nests and that urine is soaked up by the materials they've brought into their shelters. When the urine evaporates, some of the sugars and other substances crystallize and create amberat – a natural cement that helps keep the middens together over long periods of time.

Detailed record

In the arid lands of the desert, sheltered within the rock, the packrat middens can last for a very long time. Some have been dated back to the Pleistocene, about 50,000 years ago, when mammoths and sabre-toothed cats still roamed these lands. This is about as intimate a picture as climate scientists and paleontologists can get of these ancient times. The nature of the plants and animal materials the packrats collected and sealed within their middens is a natural history of how the local environment changed through seasons, years, decades and centuries.

Researchers have been studying the middens and trying to discover their secrets for decades. The United States Geological Survey even hosts a database collecting data from various packrat middens from North America. It's a rich source of detail, open to many different kinds of interpretation. A botanist might look at the species of plant in different midden layers to see how the local flora changed and what that says about climate shifts. A biologist might look at the size of the packrat dung – a proxy for the size of the packrat – and see how rodent body size has changed over time. A climate scientist might look to isotopes of carbon, oxygen and nitrogen within the collected plant material to study climate variations during the Ice Age.

Opposite A packrat of the species *Neotoma cinerea.*

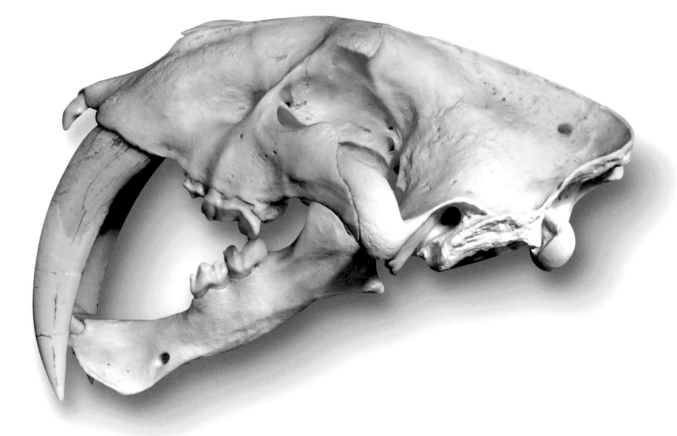

DNA evidence

Some of the most recent efforts to understand the secrets of packrat middens have focused on preserved DNA. The middens are so well preserved by the dry desert conditions that researchers are able to extract snippets of DNA and gain clues about which animals were around and when. The DNA of giant ground sloths has been found in packrat middens, for example, and some biologists have tracked the prevalence of certain diseases among packrats through time.

The middens preserve a lot of fine detail that is hard to get in the fossil record. In a given fossil site, there is often an effect known as time-averaging. It isn't always clear whether a deposit represents an hour, a month, a season, a year or some other amount of time, especially if floodwaters wash remains and organic material together into one deposit. This can obscure what we know about the past, perhaps bringing together in death organisms that did not interact in life. For instance, multiple species of tree might be represented in a fossil site, making them seem like they coexisted when they really represent a succession of different species over a century.

Tracking changes

The higher definition offered by packrat middens allows researchers to cut through some of this uncertainty, tracking how populations of particular plants and animals might encroach, then disappear, then come back, representing finer-scale environmental changes that might otherwise be missed. Packrat middens have documented how a species of juniper encroached into Montana during a dry climate, retreated to Wyoming when conditions got wet, then spread back into Montana when the climate dried again, providing a picture of how climate shifts affect organisms on timescales we can't directly see. Even though they don't know it, packrats have recorded the past and can help us to predict our future.

Above A skull of *Smilodon californicus,* a sabre-toothed cat that lived in the Americas until 10,000 years ago.

Opposite Bottom A packrat midden under a tree in Arizona.

Big Bone Lick

Extinction is a fact of life. Every species that evolves eventually disappears, never to return. But as late as the eighteenth century, this idea was controversial among naturalists. Even though humans had already driven some species, like the dodo, to extinction, experts were still uncertain whether there could be any force that totally extinguished an entire group of animals. It took some Ice Age fossils from what would become Kentucky, USA, to change people's minds.

Just a few kilometres east of the Ohio River rests Big Bone Lick. The place has carried this name for centuries – a natural salt lick for wild game that is positively brimming with the bones of Ice Age mammals. The bones of mastodons, horses, musk ox, giant sloths, tapirs and other Pleistocene beasts are abundant in this place. Many of them died after getting stuck in soft, marshy ground that created a natural trap.

Chance discoveries

The discovery that would bring this massive bonebed to the attention of naturalists came in the summer of 1739. An army of French Canadians and indigenous allies were making their way south towards the Gulf Coast in what's now known as the Chickasaw Campaign of 1739. On the way, the group stopped along the Ohio River and a party of indigenous warriors went off into the surrounding area to hunt for game. But meat wasn't all they found. The warriors returned with an immense thigh bone, tusks and several fist-sized molars.

The fossils eventually made their way to France. There, in 1762, the naturalist Louis Daubenton asserted that the bones must have come from a huge hippo and an elephant that still lived somewhere in North America. Daubenton's claims caught the attention of American naturalist and politician Thomas Jefferson. During his presidency decades later, Jefferson instructed explorers William Clark and Meriwether Lewis to stop at Big Bone Lick in 1803 and 1807 to examine the fossils and send some back east. After a failed first expedition, nearly 300 fossils made it east to the White House in 1807. The American president suspected that the creatures these bones belonged to must still live somewhere in the continent's interior, providing proof that the Americas were just as diverse and vibrant as any place in the Old World.

Extinct species

Jefferson's vision of mastodons roaming the west had already been deemed unlikely years before by a young French naturalist named Georges Cuvier. After examining large mammal bones from Europe and Siberia, Cuvier took a look at the 'animal de l'Ohio' previously described by Daubenton. He determined that these ancient bones belonged to two extinct species of elephant found nowhere on the planet today.

The fossils from Europe and Siberia came from what would eventually be called mammoths, while the bumpy molars of the

Above A 10,000-year-old molar from the extinct American mastodon, showing the distinctive rounded bumps that give this close relative of the elephant its name.

Ohio beast indicated it was a distinct species. To think that such huge animals might survive undetected – especially after centuries of global exploration – was far-fetched. These elephants were extinct. A few years later, in 1806, Cuvier more formally described the bones from Ohio. These were from a *Mastodonte*, or 'breast tooth', so-named for the rounded bumps on the molars.

Hunted to extinction?

The indigenous peoples of the Ice Age knew about these mammals. Stone tools made by their cultures are found at Big Bone Lick, and it's possible that this marshy area used to be a prime hunting ground just as it was during the French and Indian Wars. Both mammoths and mastodons are found at Big Bone Lick, as well as other large herbivores, dated to between 19,000 and 10,000 years ago. Humans were still relatively new to North America during this time, and some researchers have long suspected that people hunted some of these large mammals into extinction. Some sites around the continent preserve the cut or bashed bones of large animals, showing that people butchered Ice Age megafauna.

Ice Age experts are not totally agreed on the causes of these extinctions. The idea that America's megafauna were somehow naive and defenceless against humans doesn't make sense, nor does the extinction of non-prey species such as dire wolves and giant sloths. The climate was also rapidly changing between 19,000 and 10,000 years ago, switching from a cold and dry global climate to one that was warmer and wetter, and these shifts may have destabilized ancient ecosystems. It's possible that both humans and climate change worked in tandem, with humans adding just enough pressure to stressed ecosystems to cause collapse. The full story has yet to be told, but thanks to those who discovered and studied the Big Bone Lick fossils, we know the great mastodon and its neighbours are never coming back.

Below A drawing of a mastodon skeleton discovered in the Hudson River Valley in 1801. It was placed on display with its tusks incorrectly mounted.

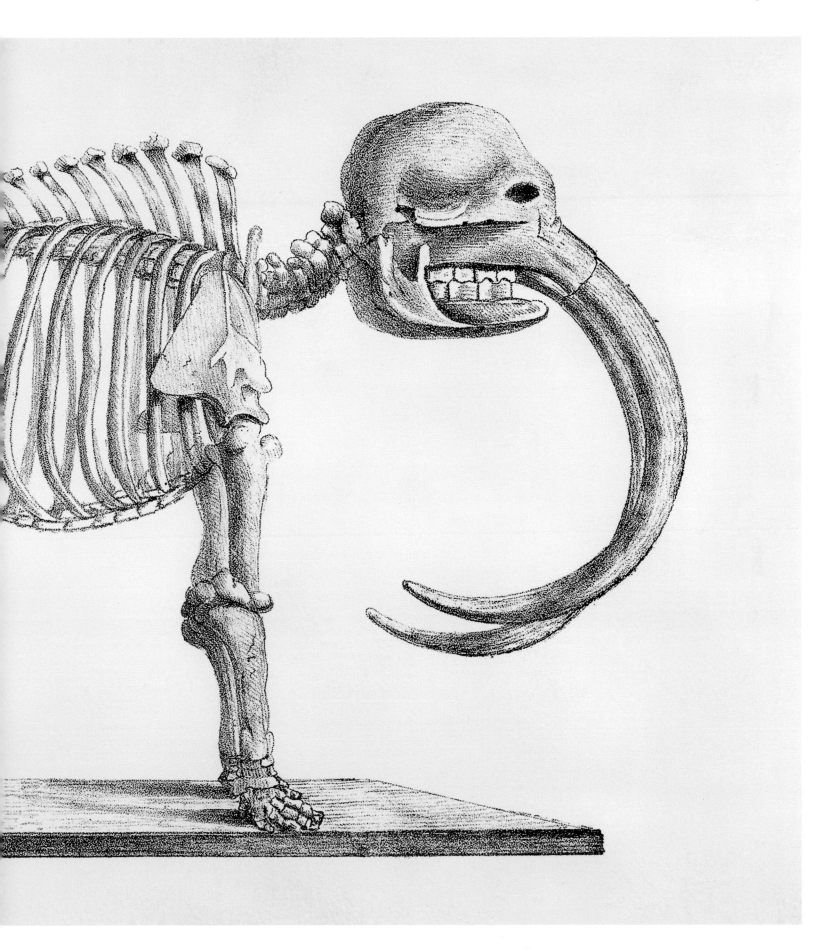

Glacial erratics

Some rocks simply look out of place. In the middle of New York City's Central Park, for example, there are immense, rough-looking rocks that have seemingly been plonked down randomly among the lawns and footpaths. These huge stones seem roughened and not exactly what a park designer would select for aesthetics. That's because they were there long before Central Park was created.

The story of Central Park's craggy boulders, as well as similar stones scattered elsewhere throughout North America, goes back to about 20,000 years ago. That was the apex of a period geologists know as the Wisconsin Glaciations, when a cooler and drier global climate allowed glaciers to expand over vast portions of the continent. Glaciers covered the entirety of what's now Greenland and Canada, and what's now New York City.

Buried by ice

Ice didn't simply crystallize over the continent in a gentle way. Pleistocene New York was buried under 300 metres of ice. All that tonnage scraped the land down to its bedrock as the ice sheet moved over the landscape, abrading and scouring the rock. Some of that stone was entirely pulverized, but some chunks of it were ploughed up and worked into the ice sheet itself. These boulders were carried along by the glacier as it slid across the land.

By about 10,000 years ago, the climate was growing warmer and wetter. This was the next pulse of a climatic back-and-forth that had gone on for hundreds of thousands of years. Colder stretches saw the expansion south of glaciers, creating steppe habitats beyond the edge of the ice where animals such as mammoths and horses roamed. During warmer swings, forests reclaimed the grasslands – the perfect habitat for mastodons, camels and giant beavers.

For the rocks, however, the change back to a warmer world meant that they would be left behind by the retreating ice. As the ice melted, the large boulders were dropped out of their frozen prisons in places far away from where they originated. This is why these rocks are called glacial erratics.

Puzzling oddities

Our present understanding of glacial erratics as geological travellers was hard-won. In the eighteenth century, during the early days of geological science, naturalists were puzzled by these large boulders that did not match the local bedrock they rested upon. The rocks had moved from someplace else, that was clear, but no one was sure of the mechanism. Some experts thought floods might be the answer. Various cultures through time had written about the power and devastation of great floods, and surely an epic flood could move a boulder as easily as a river moves a pebble.

But not everyone was sold on the idea that these rocks were transported by liquid water. Other experts countered that ice and glaciers could shape the Earth in powerful ways. Some of these large stones showed striations, or scratch marks, from being dragged

Opposite Top Glacial erratics in the Gibbon River, Yellowstone, USA.

Opposite Bottom The Great Stone of Fourstones, England.

along rather than tumbled. Other features of landscapes, such as moraines, could also be better explained by the action of ice. In time, this became the preferred explanation – when ice moves over the planet, it leaves tell-tale debris behind.

Glacial erratics aren't confined to the relatively recent deposits of the Pleistocene. In Australia's Hallet Cove Conservation Park, for example, are glacial erratics made of basalt that were left behind during the Permian Period, about 270 million years ago. Still, most glacial erratics visible in the Northern Hemisphere today are left over from the Pleistocene, and they've come to carry their own form of history.

Meaningful stones

The Cloch Chluain Fionnlocha – or Clonfinlough Stone – is found near Kinnitty in central Ireland. The stone is covered in carvings – crosses, feet, a circle divided in two, and more. Exactly when the carvings were made isn't entirely clear, but archaeologists suspect that they were made sometime during the Bronze Age, between 4,500 and 2,500 years ago. The Great Stone of Fourstones, by contrast, doesn't so much have decorative carvings as a useful staircase. This glacial erratic, located between North Yorkshire and Lancashire in England, has 15 steps carved into it to help people get to the top. In Lithuania, a glacial erratic called Puntukas is a national symbol and is festooned with images and quotes of Lithuanian pilots Steponas Darius and Stasys Girėnas to honour the tenth anniversary of their deaths while attempting to fly across the Atlantic in 1933.

Glacial erratics have taken on many different meanings and values for people through time, whether it's the Merton Stone of Norfolk, the largest erratic in England, or Plymouth Rock in Massachusetts, which traditionally marks the landing point of the *Mayflower*. But through the lens of Deep Time, they represent dramatic change. Each stone is a reminder of the ice that once gripped the world and may again in the future.

Above Plymouth Rock, marked with the year of the landing of the *Mayflower*.

Left Glacial erratics in Yorkshire, England.

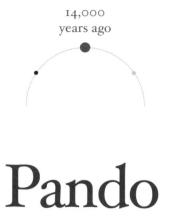

Pando

In the craggy mountains of southeastern Utah, within Fishlake National Forest, dwells what may be the largest living thing ever known. Scientists and hikers alike call this immense organism Pando.

Pando is a quaking aspen. Its scientific name, *Populus tremuloides*, immediately tells you about the character of this common tree. Quaking aspens are tall, small-diameter trees that are recognizable by their white bark and the fact that their heart-shaped leaves seem to tremble in the slightest breeze. Stand among them long enough and it feels like the forest is giving you a round of applause.

Quaking aspens aren't like most trees you might be familiar with. If you're fortunate enough to take a walk through a stand of quaking aspen, you might think that you're seeing a series of individual, discrete trees tightly packed together. But that's not the case at all. You are actually *within* a single organism, a giant that only comes into focus when you know something about the way aspens grow.

Clonal stands

Many stands of quaking aspen, like Pando, are clonal. That means that each individual tree that you see above ground is a genetically identical clone that is interlocked to its sibling clones through an underground root system. Clones can be either male or female – Pando is a huge series of male clones. New clones usually grow from root sprouts rather than seeds. What we see above ground is merely a clue to the much larger biomass beneath.

Detecting clonal stands of aspen can be a little difficult in the warm months, when their leaves are a vibrant green. But as the autumn cold starts to creep into their favoured mountain habitats, the colonies turn slightly different shades of yellow and orange at different times. This is when you can look at a mountainside, even

from far away, and get some idea of just how far these colonies can spread. But none comes close in size to Pando. First formally identified in 1976 by ecologists Jerry Kemperman and Burton Barnes, Pando has been claimed as the world's largest organism by both weight and the physical space that the tree covers. While there is no scale that could possibly weigh Pando, estimates based upon the extent of the aspen colony reach nearly 6 million kilograms – 6,000 tonnes – with a physical extent of over 40 hectares.

Ascertaining Pando's immensity wasn't just a matter of catching the colony's autumn colours. Aspen clones are genetically identical. Take a sample from one tree and the readout will be exactly the same as another in the colony. By tracking genes, biologists have been able to track Pando's size as the colony shifts with each season.

Organisms like this don't just pop up overnight. Some experts have proposed that some immense aspen stands may be as old as a million years, which means they have been around since the time mammoths, giant ground sloths, sabre-toothed cats and hyenas roamed North America. Pando is estimated to be somewhat younger than this, but it is still probably about 14,000 years old, making it both the largest and the oldest organism on Earth. While the trees on the surface only live for about a century each, the root system underground is much older and continues to grow to this day. Pando is a remnant from the end of the last Ice Age, history running through its very roots.

Opposite Every tree in the clonal colony shares the same root system.

Above Clonal stands of aspen, on a hillside in central Utah, in autumn.

Doggerland

Once upon a time, about 10,000 years ago, you could have walked from what is now mainland Europe to England. The area now covered by the English Channel was dry, part of a lost land that anthropologists and geologists know as Doggerland.

Scientists have pondered what Doggerland must have been like for over a century. HG Wells references the area in his 1897 yarn *A Story of the Stone Age*, imagining being about to walk 'dryshod' from ancient France to England, while early twentieth-century researchers studied the few plants, animal bones and tools that had been washed ashore. There had to be more, and a chance discovery by the trawler *Colinda* only intensified archaeologists' interest. In 1931, the *Colinda* brought up a large piece of peat that included a barbed antler point – a refined tool obviously made by someone who once lived on land that water later subsumed.

Along the seabed rest the remains of lions, mammoths, prehistoric people, human-made tools and ancient plants, each part of the picture of what Doggerland used to be like. It is a history that takes us hundreds of thousands of years into the past and involves several kinds of human.

Human species

Around 800,000 years ago, Doggerland hosted *Homo antecessor* – an ancient human species that, as far as we know, did not leave any evolutionary descendants. Much later, about 100,000 years ago, it was home to Neanderthals. The presence of our sister species is revealed by hundreds of tools and one piece of skull. These hunters were likely after the mammoths and woolly rhinoceros that lived here. Contrary to their brutish reputation, Neanderthals were very clever. A Neanderthal hand axe recovered from Doggerland sediments has a glob of tar on one end to make a sort of handle to better grip the tool. By about 45,000 years ago, however, the Neanderthals were gone and *Homo sapiens* had moved in. Even then, though, this landscape was not always welcoming. Conditions turned

Above The extent of Doggerland shrank from the light area 18,000 years ago to the area shaded dark green 9,000 years ago.

Opposite Ancient tree stumps exposed on a Norfolk beach by a storm in 2013.

cold about 20,000 years ago until about 15,000 years ago. People vacated Doggerland, only to return when conditions warmed.

During this last period of occupation by *Homo sapiens*, Doggerland was a relatively wet place where gentle hills rose above sopping marshes and swamps. The great glaciations late in the last Ice Age had held so much water that sea levels dropped about 70 metres lower than they are today, allowing places like Doggerland to exist. Anthropologists sometimes say that Doggerland was 'idyllic' during this time. But as the world's glaciers melted, the water levels began to rise. Each century, the waters rose by between one and two metres, gradually shrinking this ancient hunting ground. The wild game that made this place so attractive started to leave, and many of the people followed them. Then tragedy struck Doggerland.

Disaster strikes

About 8,000 years ago a series of underwater landslips along Norway's continental shelf, known as the Storegga Slides, created tsunamis. These huge waves raced towards Doggerland, reaching 40 kilometres inland and submerging much of what still remained above the water. Only a few spots and spits of land were left behind, creating a small archipelago that would eventually be covered by the rising sea levels.

As far as archaeologists have been able to discern, there were still some people living on Doggerland when the tsunamis struck. There would have been little that they could have done to protect themselves from waves four to five metres high. If there were any

survivors, they didn't stay. Even though it would have been possible to hop between the small spots of land that remained, the tsunamis effectively divided Britain from the rest of Europe for the millennia to follow.

Vast lands

It is only recently that experts have been able to understand the extent and importance of Doggerland. Relying on seismic surveys conducted by oil companies, geologists estimate that Doggerland once extended over about 180,000 square kilometres, or four times the land area of the Netherlands. Additional studies focused on mapping the area have turned up evidence of the river valleys and hills that once made it such a comfortable home.

Research and discoveries are ongoing. While trawlers often bring up unexpected relics from Doggerland, many other bones and artefacts wash up on the beaches. And while the waters here are too murky, turbulent and ship-crossed for divers to get a clear look at what's preserved below, researchers have used an array of new technologies such as radiocarbon dating and DNA analysis to better understand the land that was subsumed beneath the waves.

Above A fragment of human skull discovered off the coast of the Netherlands in 2019.

Opposite Megafauna on Doggerland included walruses, mammoths and cave lions.

Methuselah

High in eastern California's White Mountains, over 3,000 metres above sea level, lives one of the oldest trees in the world. Known as Methuselah after the long-lived Biblical patriarch, this bristlecone pine has been alive for more than 4,852 years – it is a few centuries older than the Great Pyramids of Giza.

While the great age of some of our planet's oldest trees comes as a surprise – like the immense aspen Pando (see page 184) – bristlecones are famous for their longevity. In fact, Methuselah belongs to a species called *Pinus longaeva*, or the long-lived pine. Bristlecones do not grow in what might be considered ideal conditions for most plants. They do not require plenty of moisture, rich soil or calm winds. On the contrary, these are trees that thrive where other species cannot. Bristlecone pines are often some of the first trees to take root in open ground vacated by fallen trees or cleared by rockslides, preferring harsh, dry mountain habitats where other plants have difficulty growing.

Taking things slow

The early part of a bristlecone's life is the only part that goes quickly. Trees like Methuselah often take hold in very rocky soil or even exposed rock surfaces made of dolomite and limestone. These habitats are often very high up, the elevation keeping warm growing seasons short. Bristlecones have adapted to take advantage of the warmer months at their own pace. Once they've taken root, bristlecones only grow about 0.25 millimetres each year. And while these trees grow needles like many other pines, bristlecones don't shed them at nearly the same rate. Look at a living bristlecone and the needles you see sprouting from its branches may be four decades old.

Bristlecones aren't exactly in a hurry to reproduce, either. This conifer produces both rust-coloured male cones and female cones that start off dark purple and are covered in spiky bristles – the namesake of the pine. When the gametes of the female cones are fertilized, they require about two years before they're ready to produce seeds and start the next generation.

Tough as stone

While nature doesn't always meet our expectations, long-lived bristlecones really do look just as twisted and gnarled as you might expect an ancient tree to look. Their wood is so dense that it's almost stone-like. These trees don't succumb to the same types of insect damage and rot that many others do. Instead, the resin-rich wood of the bristlecones tends to be weathered and eroded by the elements in much the same way that their anchoring stone is. Slow growth combined with extreme survival skills have produced a species of tree that can potentially grow for millennia. Methuselah is not a fluke. Many bristlecones cling on to their rocky ground for century after century.

Methuselah's great age was discovered in 1957. Then, in 1964, US Forest Service personnel cut down a bristlecone pine in Nevada that

Opposite A bristlecone growing on a rocky hillside in Inyo National Forest, California.

was 4,862 years old. It was given the posthumous title Prometheus after the titan of Greek mythology who gave fire to humanity. The accidental felling of this tree spurred a movement to better protect bristlecones from these kinds of mishaps. Bristlecones that are even older have been rumoured and await confirmation. The exact location of these trees is often undisclosed. Visitors can see the forest that Methuselah is part of, for example, but the exact tree is not pointed out for fear that tourists might try to take pieces of the it, leave graffiti or otherwise damage this forest sentinel.

A record of time

The importance of these trees goes beyond a tangible expression of time. Groves of bristlecones overlap in time with each other. By taking cores from individual trees and comparing them, matching up the chronologies and growing seasons, dendrologists have been able to create records of the past going back almost 10,000 years. The width of each ring depends on how harsh or helpful conditions were to the trees' growth, giving climate scientists an idea of climate patterns over thousands of years directly from the trees.

Archaeologists also have reason to thank trees like Methuselah. While radiometric dating techniques are often used on rock samples going back millions of years, more recent finds – from mammoths to human-made artefacts – are often dated using radiocarbon techniques. These techniques work in a similar way, just using carbon isotopes, and are great for dating samples less than 50,000 years old. When researchers had to calibrate their radiocarbon dating techniques to ensure accuracy, they turned to the bristlecones – they could count the years recorded in the rings and then take carbon-14 readings from precise dates. What was recorded in the bristlecones became the baseline for comparison, the trees themselves keeping the world's time.

Top A close-up of the tough bark of a bristlecone.

Left Methuselah Grove, California, where Methuselah is located. The tree's exact identity is kept secret.

Wrangel Island

Woolly mammoths are icons of the Ice Age. From the time these shaggy pachyderms evolved from their steppe mammoth predecessors in Eurasia about 700,000 years ago, woolly mammoths spread far and wide across the Northern Hemisphere. During their heyday, they trundled through snowy grasslands from the coast of ancient Spain to the Great Lakes region of the United States. Then, about 4,000 years ago, the last mammoths perished on a lonely Siberian island.

Known to scientists as *Mammuthus primigenius*, the woolly mammoth was one of many big beasts that thrived during the Pleistocene. The world absolutely brimmed with animals known as megafauna, defined as mammals that weigh more than 45 kilograms. Woolly mammoths certainly fit the category. While not the largest mammoth species of all time, they could still reach 6 tonnes in weight and measure more than 3 metres at the shoulder.

Out on the Ice Age landscape, woolly mammoths thrived during the colder glacial periods when the ice expanded and chilled global temperatures. Paleontologists have discerned this not only from the places mammoths have been found, but also from their teeth and gut contents. Woolly mammoths only had two kinds of teeth in their mouths – the immense curving tusks, which were modified incisor teeth, and flat, wrinkled molars. These molars look similar to those of today's elephants and best fit the profile of a grazing animal. Mammoths reached down to the ground to collect coarse grasses and other plant material, pulling the roughage up with their trunks to smash it with their flat teeth. These teeth were very different from the rough, lumpy molars of the American mastodon – a more ancient and distantly related form of Ice Age elephant – which preferred the warm, wet swamps and forests of the interglacials.

Gut contents and mammoth dung have acted as a check on what paleontologists expect from anatomy. A mammoth's menu varied from place to place and time of year, but they often lived in habitats where there were few trees. Prehistoric dung from one particular specimen, called the Yukagir mammoth, showed that, shortly

Above A woolly mammoth molar.

Opposite An artist's impression of a woolly mammoth.

before it died, this animal ate grasses, mosses, willows and even the fungus-rich dung of another mammoth.

What caused their decline?

Over time, the habitat available to mammoths shrank. Paleontologists and archaeologists still debate how this happened. Sometimes, paleontologists find mammoth bones with stone points embedded in them, hinting that ancient people may have tried to hunt them. In what's now Russia, archaeologists have found a dwelling made out of about 60 mammoth skulls – though it's not clear whether those skulls were from prey or were taken from already-dead mammoths. Still, mammoths reproduced relatively slowly, and some experts hypothesize that hungry humans hunted mammoths at rates faster than was sustainable.

Then again, around 12,000 years ago, the world went through a warming pulse called the Younger Dryas, which changed the global climate from relatively dry and cool to warmer and wetter. During these times, sea levels rose, inundating previous habitats and forcing mammoths to follow their favoured chilly tundra north as glaciers receded. The changing climate, as well as hunting by humans, might have been too much for the mammoths. That was bad news for ecosystems as a whole. Mammoths were ecosystem engineers. They pushed over trees to keep grasslands open, dispersed seeds with their dung, and provided prey for animals like sabre-toothed cats that had evolved to take on big game. When mammoths began to die back, it set off ecological ripples that could not be reversed.

Island refuge

Mammoths held on for much longer than many of their Pleistocene neighbours. While the likes of dire wolves and giant camels were gone by about 8,000 years ago, the last woolly mammoths persisted at a location known as Wrangel Island. This cold place sits in the Arctic Ocean over the edge of coastal Siberia, not all that far from Alaska, and was the last refuge of the mammoths. As far as paleontologists have been able to tell, the woolly mammoths of Wrangel Island lived there for 5,000 years longer than mammoths on the mainland before they disappeared entirely 4,000 years ago.

What killed the last mammoths? Humans have been suspected, but no evidence of hunting has been found. Through studying mammoth genetics, ancient DNA experts have found that Wrangel Island's last mammoths were going through a genetic meltdown due to inbreeding from their island isolation. Some of these mammoths may have even had shiny, translucent coats associated with this genetic bottleneck, detrimental mutations making them more vulnerable to extinction. The world had simply run out of room for these Pleistocene giants.

Above A mammoth tusk found in a riverbed on Wrangel Island.

Opposite Today, Wrangel Island is part of the Chukotka Autonomous Okrug, Russia.

3,200
years ago

Giant sequoia

Time has many different expressions. There's the digital readout of a watch, the hands of a clock, the shadow of a sundial, our concepts of today and tomorrow, the ancient-looking bones of a dinosaur, and so much else. But there are few manifestations of time as sobering and beautiful as slices of giant sequoia.

*S*equoiadendron giganteum* are the largest trees on Earth. The tallest shoot up in towers of red wood and spikey leaves over 94 metres tall, with widths reaching over 17 metres. They're so big that, before people knew better, some such trees were hollowed out so that motorists could drive right through the tree tunnels as they puttered along the Pacific Northwest. And like so many impressive trees such as the ancient *Metasequoia* and the bristlecone pines, these trees take a very, very long time to grow.

Museums around the world have slices of giant sequoia on display. These are truly immense slices of time, dwarfing the visitors that stand before them. Many were already giants long before the museums that host them were even built. For example, the slice of giant sequoia on displays at London's Natural History Museum is taken from a tree that was 1,300 years old when it was cut down in 1891.

Dating the trees

Dendrologists were able to arrive at this figure by counting the rings of the giant sequoia from the outside to the centre. The lighter-coloured bands inside a cut tree trunk are from the tree's growth periods, the lighter wood owing its shade to the larger cells created during this time. These aren't just important for determining the tree's age; they can also record whether that year was a good or bad growing season for the tree. This is useful information for investigating

Above The slice of giant sequoia on display at London's Natural History Museum is 5 metres in diameter.

Opposite A living specimen in Sequoia National Park, California.

climate, the arrival of invasive insect species, annual rainfall and other phenomena. The darker bands within the tree, by contrast, are from when the tree ended its growing season.

Sometimes it is simple enough to obtain the age of the tree – count the rings and that's the age. This might be true for trees that lived relatively comfortable lives in temperate forests where there are strong differences between the warm and cold seasons. However, droughts, diseases and even something as simple as the availability of sunlight complicate the picture. In a given year, a drought may cause a tree to stop growth early, then grow again when a nourishing rainstorm passes through, then stop growth again when dry conditions return, creating two rings for the year instead of one. A great deal of the science of dendrochronology (tree-ring dating) involves learning how to read exactly what the details of tree rings are expressing to get more accurate counts.

While it is not the longest-lived of all trees, the giant sequoia is still hardier than most. The oldest known giant sequoia is found in the Converse Basin Grove of Giant Sequoia National Monument and is over 3,266 years old. It would be a mistake to think that the oldest giant sequoia are the largest, however. Similar to us, the stature of these immense trees comes from the vigour of their adolescent growth spurts. The largest trees are the ones that grew most rapidly during their younger years, adding just a little bit more with each season as they grow ever older.

Cutting down the giants

Long-lived giants are harder to come by these days. When colonizers pushed through America's west to the Pacific in the nineteenth century, they were absolutely stunned by the size of the giant sequoias. That such trees existed was not common knowledge. People were shocked that any tree could grow so tall, and stands of giant sequoias soon became tourist attractions. Whereas local indigenous people such as the Tule River Tribe respected the trees and only used the wood when trees were felled by natural causes, the newcomers started to cut down the trees and even made dance floors of the massive stumps.

The giant sequoia slice at the Natural History Museum records these changes with a timeline next to the conserved circle of wood. When the tree that gave this record was cut down, about 70 per cent of the world's land was clothed in forest. A century later, less than 40 per cent of the world's land is forested. The accumulated growth and time that has been cut down is incalculable. The oldest trees we know of are those that are still standing, but perhaps even older giants were felled before we even knew to appreciate them.

We look to these vestiges of the past to better understand ourselves. Each piece of history – whether it's a sequoia slice or the banded iron of an ancient stone – offers a chance to reflect on how we relate to that piece. Each fragment of Deep Time is an invitation to compare our own moment to that marker, part of a still-unfolding story that gets deeper by the moment.

Above Men logging a giant sequoia in 1950.

Opposite Pioneer Cabin, a giant sequoia in California with a tunnel bored through its base, pictured here in 1865. The tree was toppled by a storm in 2017.

Hadrian's Wall

Running 117.5 kilometres from end to end, Hadrian's Wall spans England from the Irish Sea to the North Sea. It is a fortification layered with time. Begun in 122 CE during the reign of the Roman emperor Hadrian, it is the largest Roman archaeological structure in England. But the wall is not only a remnant from centuries past. During the early days of geology, it inspired an argument about the antiquity of the Earth.

The eighteenth century was a formative time for science in Europe. Many disciplines that are well known today, such as paleontology and geology, were just getting their start. By the end of the century, experts would begin to accept the reality of extinction and fossils of strange animals like the seagoing lizard *Mosasaurus* and the giant ground sloth *Megatherium* would start to fuel ideas that perhaps life had changed significantly over time. But before the importance of those discussions could truly sink in, experts had to settle a more basic question – how old was the Earth?

Biblical dating

During the seventeenth century, Irish theologian James Ussher published his estimate for the age of the Earth. Working through the genealogy of the Old Testament and drawing from other Biblical dates, Ussher proposed that the entire universe was created at 6 PM on 23 October, 4004 BCE, or around 5,600 years before Ussher's own time. English theologian John Lightfoot came up with a similar figure, and so many scholars accepted that the world must be relatively young. To these academics, there was no history outside what was recorded in the Bible.

But James Hutton thought differently. This eighteenth-century Scottish naturalist paid close attention to the various exposed rock formations he saw during his journeys. Many had the hallmarks of stones that formed in seas, much like the strata that Steno's shark teeth had come from (see page 80). It didn't make sense for rocks precipitated out of oceans to form on land. What was land had once been sea, indicating that the Earth was a great deal older than Ussher's chronology.

Hutton laid out his argument in several steps. If many rocks that were now land had originally formed as part of ancient seas, then, as Hutton wrote, "the land on which we rest is not simple and original". Following from that, Hutton proposed that the same kinds of geological processes that take place today had been in action in the past. If these things were true, then today's land had originally formed in the seas and later shifted.

Constant processes

What Hutton was proposing was one of the keys to understanding Deep Time – uniformitarianism. Geologists pay tribute to this today with the phrase 'The present is the key to the past'. In more practical terms, the idea means that the erosion, sediment deposition, volcanic activity and other geological forces that we can

Opposite The ruins of Housesteads Roman Fort, Veronicum, on Hadrian's Wall.

see today were active in the past as well. It's a powerful rhetorical point that any rock layer can be understood by applying what we can observe around us today. Everything from ancient shells on a hill to the massive bonebeds of Jurassic dinosaurs can be investigated and understood through the lens of processes we can still observe.

But Hutton faced an uphill battle. What's always been taken as true cannot be overturned in an instant. He required proof that Earth's strata were very old and needed vast amounts of time to form, move, and come to their present positions. Hadrian's Wall formed one of his key lines of evidence.

Unchanging wall

Even in Hutton's time, experts knew that Hadrian's Wall was built during the Roman occupation of the island some 2,000 years earlier. The wall was largely made from stone, particularly limestone. For the traditional biblical chronologies to be accurate, the wall would have to have undergone some truly massive changes. After all, the same amount of time was supposed to account for phenomena like seabeds being deposited high on land. But that's not what the geologic signature of Hadrian's Wall revealed. The stone of the wall had hardly changed since the time of its construction, meaning that processes like erosion must play out over time frames that few had previously comprehended.

Part of what made Hutton's example so ingenious is that a date for Hadrian's Wall was known and accepted. Even today, it can be difficult to know how long it took for a particular rock layer to accumulate, while various forms of weathering can occur at varying – albeit still slow – rates. Had Hutton picked just any rock strata, critics might have countered that no one could know the absolute age of such rocks nor their complete history. A wall made of stone circumvented that argument, adding a key piece of evidence to the world-changing ideas that Hutton proposed.

Above A view from Walltown Crags on Hadrian's Wall.

Glossary

Angiosperms

Plants that produce flowers and seeds within a fruit. Also called flowering plants, angiosperms first appeared at least 125 mya. They have since diversified to become the most diverse group of plants, with more than 300,000 known species.

Archaean Eon

A geological eon that lasted from 3.9 bya to 2.5 bya. During the Archaean, the Earth cooled sufficiently for the first continents to form.

Archosaur

A clade of tetrapods that includes dinosaurs, pterosaurs, crocodilians and birds. The first archosaurs appeared in the Early Triassic, 250 mya.

Baryon

A subatomic particle made from at least three quarks, such as a proton or a neutron.

Basalt

A porous igneous rock that forms from the rapid cooling of lava at or near the surface of a rocky planet or moon. Basalt makes up more than 90 per cent of all volcanic rock on the Earth.

Big Bang

The moment at which space and time are theorized to have come into existence as a singularity. Detailed study of fluctuations in the CMBR has led to an estimate that the Big Bang occurred 13.77 billion years ago.

Biostratigraphy

The dating and correlating of rock strata according to the fossils found in them.

Breccia

Rock formed of small fragments that have been cemented together.

Cambrian Period

A geological period that lasted from the end of the Ediacaran, 541 mya, to the start of the Ordovician, 485 mya. The start of the Cambrian saw a rapid diversification of life forms known as the Cambrian Explosion. This lasted for between 13 and 15 million years, during which time nearly all the major animal phyla make their first appearance in the fossil record.

Carboniferous Period

A geological period that lasted from the end of the Devonian, 359 mya, to the start of the Permian, 299 mya. Coal seams are formed from the remains of the vast forests of the Carboniferous, a period during which oxygen levels in the air were particularly high, allowing terrestrial invertebrates to grow to great sizes.

Chemosynthesis

A process by which bacteria and other organisms make food using the energy released by inorganic chemical reactions. Chemosynthesis allows life to exist in places where there is no light available for photosynthesis, such as around deep-sea hydrothermal vents.

Cosmic Microwave Background Radiation (CMBR)

A faint radiation that fills space. It was emitted 380,000 years after the Big Bang, at the moment when the universe first became transparent to radiation. The CMBR is the oldest light in the universe.

Cretaceous Period

A geological period that lasted from the end of the Jurassic, 145 mya, to the start of the Paleogene, 66 mya. The Cretaceous ended abruptly with the Cretaceous–Paleogene (K–Pg) extinction event. The last of the five mass extinctions, it was caused by an asteroid strike and killed off the non-avian dinosaurs.

Dendrology

The scientific study of wooded plants (trees, shrubs and lianas).

Devonian Period

A geological period that lasted from the end of the Silurian, 419 mya, to the start of the Carboniferous, 359 mya.

Dinosaur

A diverse clade of reptiles that first appeared between 243 and 233 mya. All non-avian dinosaurs became extinct 66 mya during the K–Pg mass extinction, but modern birds continue the dinosaur lineage.

Ediacaran Period

A geological period that lasted from the end of the Cryogenian, 635 mya, to the start of the Cambrian, 541 mya.

Endosymbiosis

A symbiotic relationship in which one organism lives inside another. The first eukaroyotic cells formed through a process of endosymbiosis, which resulted in the formation of the organelles mitochondria and chloroplasts.

Entropy

In physics, a measure of the disorder of a closed system. The higher a system's entropy, the more disordered it is said to be. The Second Law of Thermodynamics states that entropy always increases over time in a closed system. The concept of entropy was first formulated in the study of steam engines, but it has recently been proposed as an explanation of our perception of the arrow of time, in which entropy increases from the past to the future.

Eocene

A geological epoch that lasted from 56 mya to 34 mya. The Eocene saw the first appearance of many modern groups of mammals.

Eukaryote

An organism whose cells have a nucleus in which the DNA is contained. The cells also contain organelles called mitochondria, which generate chemical energy using the molecule adenosine triphosphate (ATP).

Glacial erratic

A rock that was carried along by a glacier and deposited far from its place of origin when the glacier melted.

Gneiss

A common form of metamorphic rock, formed when igneous or sedimentary rocks are placed under high temperature and pressure.

Gondwana

A huge landmass that formed about 550 mya. About 335 mya, Gondwana merged with the landmass Laurussia to form the supercontinent Pangaea.

Gymnosperm

A group of seed-producing plants that includes conifers and cycads. The name means 'naked seed', contrasting gymnosperms with flowering plants in which the seed is enclosed within an ovary. Gymnosperms first appeared during the Carboniferous, about 319 mya.

Hadean Eon

A geological eon that lasted from 4.49 bya to 3.9 bya. Most of the Earth's surface was made up of molten rock during the Hadean Eon.

Hominin

The group of bipedal apes to which modern humans belong. Hominins share a last common ancestor with chimpanzees and bonobos about 6 million years ago. Modern humans are the only surviving species of hominin.

Ichthyosaurs

A group of large, fish-like marine reptiles that lived between 250 and 90 mya.

Inflation

Also called cosmic inflation, an epoch that is thought to have lasted from 10^{-36} seconds to 10^{-32} seconds after the Big Bang, during which time the size of the universe doubled in size at least 85 times.

Jurassic Period

A geological period that lasted from the end of the Triassic, 201 mya, to the start of the Cretaceous, 145 mya.

Laurussia

A northern landmass that merged with Gondwana 335 mya to form Pangaea.

Light year

The distance travelled by light through a vacuum in one year. It is equal to 9.46 trillion kilometres.

Megafauna

A loosely defined term applied to large animals, usually those weighing more than 45 kilograms. The Pleistocene megafauna were large mammals, such as mammoths, mastodons and giant sloths, that became extinct during the Late Pleistocene.

Meteorite

A rock that falls to the Earth from space. Most meteorites are fragments of shattered asteroids, but they can also come from planets, comets or the Moon.

Microfossil

The tiny remains of microorganisms such as bacteria, protists and fungi.

Neanderthal

An extinct species of human that lived in Eurasia until about 40,000 years ago. Neanderthals interbred with modern humans, and Neanderthal DNA is found in all non-African human populations.

Neutrino

A subatomic particle with no electric charge and a very small mass. Neutrinos are among the most abundant particles in the universe, but they are very hard to detect as they have very little interaction with other matter.

Ordovician Period

A geological period that lasted from the end of the Cambrian, 485 mya, to the start of the Silurian, 444 mya. The first of the five mass extinctions occurred at the end of the Ordovician.

Ornithiscian

A clade of mostly herbivorous dinosaurs. Although the name means 'bird-hipped', ornithiscians are only distantly related to modern birds, which are theropods.

Orogeny

The structural deformation of the Earth's crust at convergent plate boundaries that causes mountain ranges to form.

Paleobotany

The scientific study of plant fossils.

Paleoproterozoic Era

A geological era that lasted from 2.5 bya to 1.6 bya. Photosynthesis increased enormously during this era, oxygenating the atmosphere. The era also saw the appearance of the first eukaryotic cells.

Paleozoic Era

A geological era lasting from 541 to 252 mya. Beginning with the Cambrian Explosion, this was an era of huge geological and biological changes on the Earth.

Pangaea

A supercontinent centred on the equator that comprised most of the Earth's landmass. Pangaea formed 335 mya and started to break apart 175 mya.

Permian Period

A geological period that lasted from the end of the Carboniferous, 299 mya, to the start of the Triassic, 252 mya. It ended with the largest mass extinction event in Earth's history.

Phagocytosis

The process by which one cell ingests another by completely surrounding it.

Plasma

A state of matter consisting of charged ions. Gas is turned into plasma by adding energy, which strips the negatively charged electrons from the positively charged nuclei. More than 99 per cent of the visible matter in the universe is in the form of plasma, mainly contained within stars.

Plate tectonics

The scientific theory that describes the movement of the plates that make up the Earth's crust. The theory explains how continents have changed over geological time.

Pleistocene

A geological epoch that lasted from 2.6 million years ago to 11,700 years ago. The Earth's most recent ice ages took place during the Pleistocene.

Precambrian

An informal unit of geological time spanning from the formation of the Earth, 4.6 bya, to the start of the Cambrian Period, 541 mya.

Prokaryotes

Single-celled organisms whose cells do not contain a nucleus or organelles. Prokaryotes are divided into two domains: bacteria and archaea.

Pseudosuchian

A clade of archosaurs that includes living crocodilians and their related ancestors.

Radiometric dating

A method for calculating the age of rocks based on the half-life of a radioactive constituent of the rock such as uranium or potassium-40. The age of the rock is determined by measuring the ratio of the radioactive element to the element into which it decays.

Redshift

An increase in the wavelength of electromagnetic radiation. It is seen when the source of light is moving away from an observer. An observed redshift in the light from distant galaxies provided the first evidence that the universe is expanding.

Selenology

The scientific study of the Moon.

Singularity

In physics, a point in space-time at which the quantities used to measure the gravitational field become infinite, physical laws are indistinguishable from one another, and space and time merge into one. The Big Bang theory postulates that the universe began as a singularity 13.77 bya. Singularities are also calculated to exist inside black holes.

Standard ruler

An astronomical object whose physical size is known, allowing its distance from Earth to be measured.

Stromatolite

A fossilized mineral deposit laid down over time by many generations of photosynthesizing cyanobacteria.

Supernova

The explosion of a star. A supernova can occur in two ways: at the end of a star's life cycle when its core collapses; or when one member of a binary star system attracts matter from its companion star, causing it to explode.

Superposition (Law of)

A basic principle in geology stating that, within a sequence of layers in sedimentary rock, the oldest rocks will be at the bottom and the youngest at the top, unless the rocks have been deformed and tilted. The Law of Superposition was first formulated in 1669 by Danish geologist Nicolas Steno.

Tektites

Gravel-sized bodies of natural glass that are formed and ejected by meteorite strikes.

Theropod

A diverse clade of dinosaurs that first appeared 231 mya and includes *Tyrannosaurus rex* and modern birds.

Triassic Period

A geological period that lasted from the end of the Permian, 252 mya, to the start of the Jurassic, 201 mya.

Trilobites

A highly successful group of marine arthropods that had an easily fossilized exoskeleton. Trilobites first appear in the fossil record 521 mya. They went extinct at the end of the Permian about 252 mya.

Uniformitarianism

In geology, the theory that changes in the Earth's crust result from the action of continuous and uniform processes, meaning that the present is key to the past.

Uranium–lead dating

A method of dating rock by measuring the proportion of uranium in the rock that has decayed into lead. This method can date rocks with an error of less than 1 per cent.

Zircon

A mineral found in igneous rocks that can be used to date the rock using uranium–lead dating.

Above Fairy chimneys in Cappadocia, Turkey. These
unusual features are formed from eroded volcanic rock.

ichthyosaurs, 105, *106–107*, 124, *126*
Iguanodon, 132–134, *133*
marine species, 104–106, *105–107*
Odontochelys, 104
Permian and Triassic periods, 102
plesiosaurus, *124–125*
rise of, 100
Tanystropheus, 104
Triassic period, 112–115
Reusch, Hans, 68–71, *68*
Reusch's Moraine (Norway), 68
rock classifications, 40
and dating, 36–37, 38

S

Shipley, John, 28
Siberian Traps (Russia), 102
Slave Crayton (Canada), *41*
Smith, William
'faunal succession' principle, 64
fossil drawings, *66–67*
Smith's Geological Map of England,
64–67, *65*
Smith's Geological Map of England,
64–67, *65*
Snider-Pellegrini, Antonio, 116
Creation and its mysteries unveiled, 116
Snider-Pellegrini Map, 116–119, *117*
Snowball Earth hypothesis, 68–71, 75
Diamictite (Virginia, USA), formation
of, *71*
glaciation, and evidence of, 68
ice-albedo feedback, *70*, 71
Sowerby, James, 28–31
space-time, 8, *9*
Sprigg, Reginald, 72
fossils found by, *72*, *73*, *75*
standard ruler length, 16, *17*

Steno, Nicolas, *81*, 83
stromatolites, *49–51*
cyanobacteria, and formation of, 48
fossilization of, 48–51
supernovae, 24–27
Kepler's Supernova, 24, *25*
Supernova 1997ff, 27, *28*
white dwarf star, 24

T

tektites, 168–171, *168*, *171*
Bolaven Plateau (Laos), impact site,
169
Topham, Edward, 28, *28*
trees
dating, 200–202
Early Cretaceous period, *137*
giant sequoia, 200–203, *200–203*
Metasequoia (dawn redwood), 140–143,
140–143
Methuselah (bristlecone pine),
192–195, *193–195*
Pando (quaking aspen), 184–187,
185–187
Triassic period, 100–102, 112–115
marine reptiles, 104–105
reptiles, and rise of, 100, 112–115
Triassic-Jurassic extinction, 112–115
trilobites
Cambrian and Ordovician Eras, 87
fossil record of, 76, *77*, 84, *85*

U

uniformitarianism, 204–205
Upheaval Dome (Utah, USA), *170*
uranium, 37

V

volcanic eruptions, 102
Emeishan Traps (China), 102
and mass extinctions, 102, 112, *113*
Pangaea, impact on, 112
Siberian Traps (Russia), 102

W

Walcott, Charles Doolittle, 76, *78*
Weald, the (southeast England), 132–135,
135
geological map, *133*
Wegener, Alfred, *116*, 119
map of Pangaea, *117*
Wold Cottage meteorite, 28–31, *29*, *31*
obelisk erected to commemorate, *30*
witness accounts, 28
woolly mammoths, 196–199, *196–199*
molar from, *196*
Wrangel Island (Siberia), remains
found at, 199
Wrangel Island (Siberia), *198*
woolly mammoth remains found at,
199

Y

Yellowknife (Canada), 40

Z

zircon crystals, 36–39, *36*
rock dating, 37, 38
uranium, 37

Index

Above The Antelope Canyon, Arizona

Above The Giant's Causeway in Northern Ireland comprises 40,000 interlocking basalt columns.
It formed during a period of intense volcanic activity between 50 and 60 million years ago.

Above The Rainbow Mountains of the Zhangye Danxia Landform, China.
The stripes are created by traces of iron and other minerals in the rock.

Credits